NASA SP-287

WHAT MADE APOLLO A SUCCESS?

A series of eight articles reprinted by permission from the March 1970 issue of Astronautics & Aeronautics, *a publication of the American Institute of Aeronautics and Astronautics*

Scientific and Technical Information Office   1971
NATIONAL AERONAUTICS AND SPACE ADMINISTRATION
Washington, D.C.

CONTENTS

Section		Page
1.	INTRODUCTION	1
	By George M. Low	
2.	DESIGN PRINCIPLES STRESSING SIMPLICITY	15
	By Kenneth S. Kleinknecht	
3.	TESTING TO ENSURE MISSION SUCCESS	21
	By Scott H. Simpkinson	
4.	APOLLO CREW PROCEDURES, SIMULATION, AND FLIGHT PLANNING	31
	By Warren J. North and C. H. Woodling	
5.	FLIGHT CONTROL IN THE APOLLO PROGRAM	41
	By Eugene F. Kranz and James Otis Covington	
6.	ACTION ON MISSION EVALUATION AND FLIGHT ANOMALIES	53
	By Donald D. Arabian	
7.	TECHNIQUES OF CONTROLLING THE TRAJECTORY	59
	By Howard W. Tindall, Jr.	
8.	FLEXIBLE YET DISCIPLINED MISSION PLANNING	69
	By C. C. Kraft, Jr., J. P. Mayer, C. R. Huss, and R. P. Parten	

TABLES

Table		Page
1-I	DEVELOPMENT AND QUALIFICATION TESTS	3
1-II	HISTORY OF ENVIRONMENTAL ACCEPTANCE TEST FAILURES	5
1-III	APOLLO FLIGHT ANOMALIES	7
3-I	LUNAR MODULE ACCEPTANCE VIBRATION TEST IN 1967	23
3-II	COMMAND AND SERVICE MODULE ACCEPTANCE VIBRATION TEST IN 1967	24
3-III	CRITERIA FOR LUNAR MODULE QUALIFICATION AND ACCEPTANCE THERMAL TESTING IN 1967	25
7-I	MANDATORY GUIDANCE, NAVIGATION, AND CONTROL SYSTEMS	
	(a) Lunar module systems	62
	(b) Command and service module systems	63
8-I	APOLLO SPACECRAFT FLIGHT HISTORY	72

FIGURES

Figure		Page
1-1	Vibration test level for acceptance	4
1-2	Thermal test level for acceptance	4
1-3	Results of vibration acceptance tests for 11 447 tests of 166 different components	5
1-4	Results of thermal acceptance tests for 3685 tests of 127 different components	5
1-5	Apollo 10 fuel-cell temperature oscillations as they originally appeared in flight	7
1-6	Disturbance of Apollo 10 fuel-cell temperature as it was identified in the laboratory	7
1-7	Buildup of Apollo mission capability	9
2-1	Mission Control Center, Houston, Texas	17
2-2	Service propulsion engine propellant valve and injector	18
2-3	Command module hatch	19
3-1	Revised Apollo acceptance vibration test guidelines	23
3-2	Revised Apollo acceptance thermal test guidelines	25
3-3	Results of acceptance vibration tests for 11 447 tests of 166 different components	26
3-4	Results of acceptance thermal tests for 3685 tests of 127 different components	26
3-5	Installation of the acceptance-tested crew equipment in the Apollo command module at the NASA Manned Spacecraft Center	27
4-1	Astronaut trains underwater in simulated zero-g condition in water-immersion facility. Astronaut wears weights on shoulders, wrists, and ankles. Total ballast is about 180 pounds	32
4-2	Lunar landing training vehicle trains crews for last 500 feet of altitude in critical moon landing phase	33

Figure		Page
4-3	Lunar module mockup installed in KC-135 aircraft. Support structure takes loads imposed in 2-1/2g pullup, after which zero g is achieved for 20 to 30 seconds on a parabolic flight path	33
4-4	Familiarization run on the mobile partial-gravity simulator used for lunar walk indoctrination	34
4-5	Apollo 12 landing and ascent model of Surveyor and Snowman craters as seen from 1800 feet	36
4-6	Command module procedures simulator and lunar module procedures simulator	36
4-7	Translation and docking trainer simulates lunar module active docking over last 100 feet of separation distance	37
4-8	Dynamics crew procedures simulator	37
4-9	Simulated command module crew station for the dynamics crew procedures simulator	37
4-10	Lunar module mission simulator with crew station and Farrand optical systems for three windows	38
4-11	Visual optics and instructor station for command module mission simulator	38
4-12	Apollo 12 landing and ascent visual simulation system	38
5-1	Electrical power display when Apollo 12 was at an altitude of 6000 feet	42
5-2	Mission-development time line	43
5-3	Mission Operations Control Room divisions	44
5-4	Partial sample of CSM systems schematic	45
5-5	Sample of flight mission rules	47
5-6	Sample of Flight Control Operations Handbook	49
5-7	Sample of Flight Controller Console Handbook	50
5-8	Sample of programed-instruction text	51
5-9	Logic of flight control decisions	52

Figure		Page
6-1	Mission evaluation room with team leaders' table in the foreground and discussion of a system problem in the background	54
6-2	Long-range photography of adapter failure during the Apollo 6 mission	56
7-1	Steps the ground-based flight controllers take if certain guidance and control values exceed premission limits for the LM during LM descent to the lunar surface	65
8-1	Iterative mission-planning process	69
8-2	Apollo mission design instrumentation	74

1. INTRODUCTION

By George M. Low
Manned Spacecraft Center

On July 20, 1969, man first set foot on another planet. This "giant leap for mankind" represented one of the greatest engineering achievements of all time. This article and the others in this document describe and discuss some of the varied tasks behind this achievement.

We will limit ourselves to those tasks that were the direct responsibility of the NASA Manned Spacecraft Center: spacecraft development, mission design and mission planning, flight crew operations, and flight operations. We will describe spacecraft design principles, the all-important spacecraft test activities, and the discipline that evolved in the control of spacecraft changes and the closeout of spacecraft anomalies; and we will discuss how we determined the best series of flights to lead to a lunar landing at the earliest possible time, how these flights were planned in detail, the techniques used in establishing flight procedures and carrying out flight operations, and, finally, crew training and simulation activities — the activities that led to a perfect flight execution by the astronauts.

In short, we will describe three of the basic ingredients of the success of Apollo: spacecraft hardware that is most reliable, flight missions that are extremely well planned and executed, and flight crews that are superbly trained and skilled. (We will not discuss two equally important aspects of Apollo — the launch vehicles and launch operations. These elements, the responsibility of the NASA Marshall Space Flight Center and the NASA Kennedy Space Center, go beyond the scope of this series of articles.)

SPACECRAFT DEVELOPMENT

Four aspects of spacecraft development stand out: design, test, control of changes, and interpretation of discrepancies. We can begin with them.

Spacecraft Design

The principles of manned spacecraft design involve a combination of aircraft-design practice and elements of missile-design technology: Build it simple and then double up on many components or systems so that if one fails the other will take over. Examples are ablative thrust chambers that do not require regenerative cooling; hypergolic propellants that do not require an ignition source; three fuel cells, where one alone could bring the spacecraft back from the moon; series/parallel redundancy in valves, regulators, capacitors, and diodes so that neither an "open" nor a "closed" failure will be catastrophic.

Another important design rule, which we have not discussed as often as we should, reads: Minimize functional interfaces between complex pieces of hardware. In this way, two organizations can work on their own hardware relatively independently. Examples in Apollo include the interfaces between the spacecraft and launch vehicle and between the command module and the lunar module. Only some 100 wires link the Saturn launch vehicle and the Apollo spacecraft, and most of these have to do with the emergency detection system. The reason that this number could not be even smaller is twofold: Redundant circuits are employed, and the electrical power always comes from the module or stage where a function is to be performed. For example, the closing of relays in the launch vehicle could, in an automatic abort mode, fire the spacecraft escape motor. But the electrical power to do this, by design, originates in the spacecraft batteries. The main point is that a single man can fully understand this interface and can cope with all the effects of a change on either side of the interface. If there had been 10 times as many wires, it probably would have taken a hundred (or a thousand?) times as many people to handle the interface.

Another design question for manned flight concerns the use of man himself. Here again, we find no simple rule as to how man should interface with his machine. Generally, tedious, repetitive tasks are best performed automatically; and selection of the best data source to use, selection of control modes, and switching between redundant systems are tasks best performed by the pilot. In Apollo, the trend has been to rely more and more on automatic modes as systems experience has been gained. For example, computer programs for rendezvous were reworked to require far less operator input than had originally been planned, but the entire rendezvous sequence was designed so that the pilot could always monitor the automatic system's performance and apply a backup solution if deviations were noted. A tremendous amount of time and effort was spent to determine how the crew could best decide which data source to use and which of many redundant systems to rely on. This was always a basic mission-design consideration.

The concept of inflight maintenance was discarded entirely as being impractical for flights with the specific purpose and duration of Apollo. In its place, more telemetry was added and full advantage was taken of the ground's ability to assess system performance, predict trends, and compare data with preflight test experience.

Apollo Test Activities

The single most important factor leading to the high degree of reliability of the Apollo spacecraft was the tremendous depth and breadth of the test activity.

There are two general categories of tests: (1) those made on a single prototype device (or on a few devices) to demonstrate that the design is proper and will perform properly in all environments and (2) those made on each flight item to assure that there are no manufacturing errors and that the item will function as intended. Both categories apply to individual parts, components, subsystems, systems, and entire spacecraft. The first category includes development testing early in the design cycle and the very formal certification or qualification tests performed on test articles identical to the flight system. The second category covers acceptance testing.

Instead of reviewing the entire development and qualification test program, we can focus on only those tests involving complete spacecraft or boilerplates, as listed in table 1-I. Each of these tests taught us more about our spacecraft — their strengths and weaknesses. As a result of the thermal vacuum tests, the spacecraft withstood the translunar and lunar environments without a single thermal problem. Passive thermal-control modes were developed that required minimum crew inputs and gave a perfect thermal balance. The land-impact tests demonstrated that the command module could survive an emergency land landing if wind velocity stayed within certain limits. These tests also led to the design of a new impact-attenuation strut for the astronaut couch. The strut allowed us to increase the permissible launch wind speed and thereby gave us more flexibility in an otherwise constrained launch window. Other tests brought equally significant results.

TABLE 1-I. - DEVELOPMENT AND QUALIFICATION TESTS

[Full-scale spacecraft testing]

Escape motor flight tests	7
Parachute drop tests	40
Command module land impact tests	48
Command module water impact tests	52
Lunar module structural drop tests	16
Lunar module complete drop tests	5
Command and service module acoustic/vibration tests, hr	15.5
Lunar module acoustic/vibration tests, hr	3.5
Command and service module modal survey testing, hr	277.6
Lunar module modal survey testing, hr	351.4
Command and service module thermal vacuum tests, hr	773
Lunar module thermal vacuum tests, hr	2652
Service module propulsion-system tests, min	1474.5
Ascent-stage propulsion-system tests, min	153
Descent-stage propulsion-system tests, min	220

Most important of all, the tests gave us a tremendous amount of time and experience on the spacecraft and their systems. Such experience — together with a detailed analysis of all previous failures, discrepancies, and anomalies — led us to the conclusion that we were ready to fly a lunar orbit with Apollo 8 and that we were ready to make a lunar landing with Apollo 11.

Acceptance testing played an equally important role. This testing starts with piece parts. Although Apollo was late in applying this rule, I believe that screened and burned-in electronic parts must be made a firm requirement. Next, each component, or black box, is tested before it is delivered, and again before it is installed in the spacecraft. Then, factory testing of the complete spacecraft begins. First, the wiring is wrung out, and individual subsystems are tested as installed. Then, groups of systems are jointly tested. Finally, the complete spacecraft, with all of its systems functioning, is run in an integrated test. All normal, emergency, and redundant modes are verified.

After delivery to the launch site, similar (when practical, identical) tests are performed. A major test at Cape Kennedy is a manned altitude-chamber run of each spacecraft. The final acceptance test, of course, is the countdown itself.

A most important facet of acceptance testing is environmental acceptance testing. The primary purpose of acceptance vibration testing and acceptance thermal testing is to find workmanship errors. To do this, the environment has to be severe enough to find the fault (e.g., a cold-solder joint), yet not so severe as to weaken or fatigue the component. Figures 1-1 and 1-2 show the levels selected for these tests in Apollo. These levels were picked on the basis of experience in Gemini and other programs. Each component type, of course, had to pass qualification tests under even more severe environments. Nevertheless, our environmental acceptance tests sometimes uncovered design faults (as opposed to workmanship faults) that had been missed in the qualification tests. The reason was that a single qualification test may have missed a marginal condition, which the large number of acceptance tests could catch.

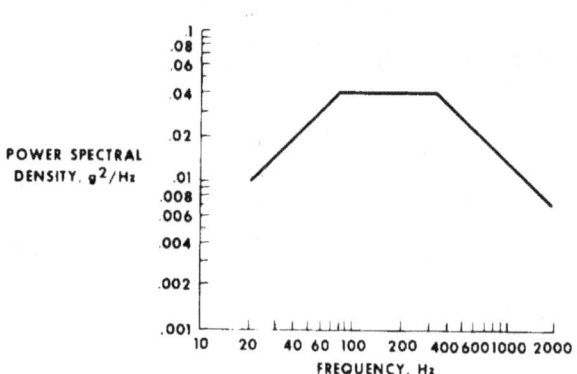

Figure 1-1. - Vibration test level for acceptance.

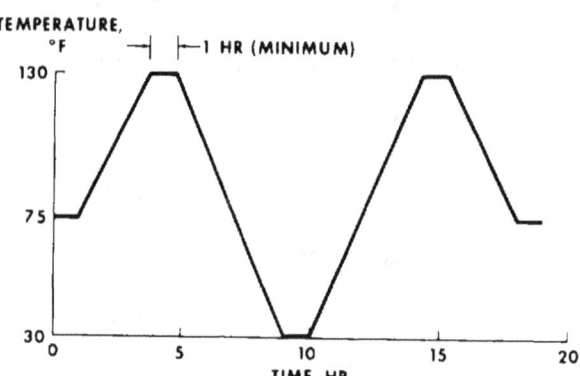

Figure 1-2. - Thermal test level for acceptance.

We also considered environmental acceptance tests of complete spacecraft, but decided against this because the environment on most components, as mounted in the spacecraft, is not severe enough to bring out workmanship faults. The vibration levels on many components are one or two orders of magnitude less than those given in figure 1-1. (This conclusion would not be true for smaller, more compact spacecraft.) Temperatures in the spacecraft generally remain constant because most electronic components are mounted on cold plates.

Figures 1-3 and 1-4 summarize the results of the Apollo environmental acceptance test program. Note that 5 percent of all components failed under vibration, and 10.3 percent of all components did not pass the thermal testing. Remember that these components were otherwise ready for installation in the spacecraft. By category, the failure modes look like those listed in table 1-II. If these tests had not been performed, and if these failures had occurred in flight, we very likely would still be waiting for the first manned lunar landing.

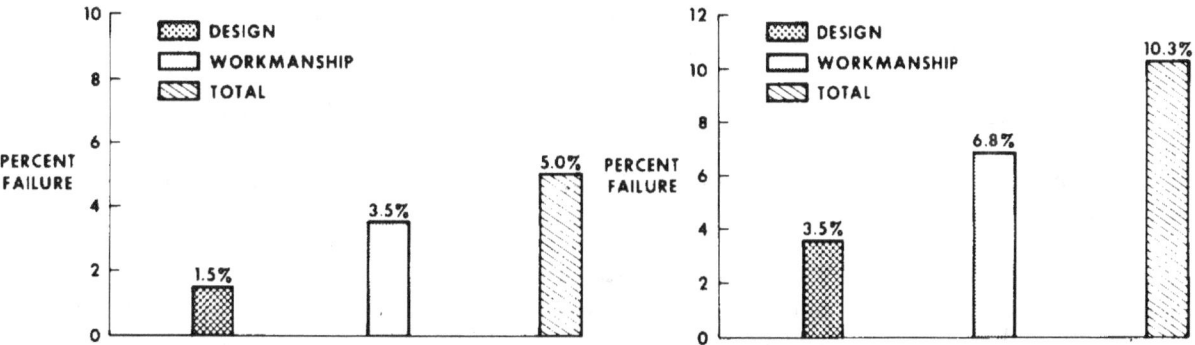

Figure 1-3. - Results of vibration acceptance tests for 11 447 tests of 166 different components.

Figure 1-4. - Results of thermal acceptance tests for 3685 tests of 127 different components.

TABLE 1-II. - HISTORY OF ENVIRONMENTAL ACCEPTANCE TEST FAILURES

Mode	Percent failed
Electrical	57.3
Mechanical	27.4
Contamination	11.5
Other	3.8
Total	100

Control of Changes

If the design has been verified and if a thorough test program has been completed, it should not be necessary to make any changes. Of course, this idealized situation does not exist in any program like Apollo where design, test, and flight often overlap and must be carried out at the same time. Changes may be required as a result of test failures, or another look at the design may identify a situation that could lead to a failure or to the inability to react to failure. Sometimes a more detailed definition of flight missions or operational use of the hardware itself leads to a requirement for change.

Since it is not possible to eliminate all changes, we have to start with the premise that any change will be undesirable. That is, a change will void all previous test and flight experience and, no matter how simple, may have ramifications far beyond those identified by the initial engineering analysis.

Because changes must be made nevertheless, it becomes important to understand and to control them, no matter how small. In Apollo, we handled all changes through a series of Configuration Control Panels and a Configuration Control Board. The panels considered minor hardware changes early in the development cycle, as well as crew procedures and all computer programs. The Board considered more significant hardware changes, all hardware changes after spacecraft delivery, and procedures or software changes that could affect schedules or missions.

The Apollo Spacecraft Configuration Control Board met 90 times between June 1967 and July 1969, considered 1697 changes, approved 1341, and rejected 356. We had a low rejection rate because proposed changes were reviewed before they came to the Board, and only those deemed mandatory for flight safety were brought before it. The Board is chaired by the Program Manager, who also makes the final decision on all changes. The Board includes the directors of all major technical elements of the NASA Manned Spacecraft Center and the contractors' program managers.

We considered changes large and small. An example of a large change is the new spacecraft hatch that was incorporated after the fire. However, we reviewed in equal technical detail a relatively small change, such as a small piece of plastic to go inside the astronaut's ballpoint pen.

The Board was established to discipline the control of changes; but it was found to serve a much larger purpose: It constituted a decisionmaking forum for spacecraft developer and user. In reaching our decisions, we had the combined inputs of key people representing hardware development, flight operations, flight crews, safety, medicine, and science.

I have recently reviewed the results of the 90 Board meetings that preceded Apollo 11. Even with hindsight, I find few, if any, Board decisions that I would make differently today.

Closeout of Failures

Throughout Apollo, many discrepancies or failures occurred daily. The relationship may have been a close one (i.e., failures actually took place during testing of the next spacecraft to fly) or it might have been remote (i.e., a component identical to one used on Apollo failed on another program). In either instance the result was the same: The failure had to be understood and, if applicable, some corrective action taken. This might involve design change, reinspection, or perhaps procedural change.

I will confine my remarks to anomalies that occurred during the first five manned Apollo flights. The number of anomalies for each mission are given in table 1-III. Note that, even though each of the flights was completely successful and met all its objectives, the number of anomalies went quite high. Perhaps this is the best proof of the validity of the Apollo design concept: The spacecraft were designed for mission success.

TABLE 1-III. - APOLLO FLIGHT ANOMALIES

Spacecraft	Number of anomalies	
	Command and service module	Lunar module
Apollo 7	22	--
Apollo 8	8	--
Apollo 9	14	12
Apollo 10	23	15
Apollo 11	9	13

The closeout of these flight failures had to be done in the time available between the completion of one flight and the start of the next — a period usually only about 6 weeks long. Yet even these 6 weeks were not fully available, because hypergolic propellants were loaded into the spacecraft a month before launch, thereafter severely limiting ability to make spacecraft changes and to perform necessary retesting. Nevertheless, each of the failures listed in table 1-III was satisfactorily closed out before the next flight.

Let us look at just one example. On Apollo 10, during several of the lunar orbits, a critical fuel-cell temperature started to oscillate significantly, as shown in figure 1-5. Normally, this temperature is steady, between 155° and 165° F. The oscillations encountered on Apollo 10 triggered the spacecraft alarm system, but otherwise were not detrimental. Yet, unless we understood their cause, we could not be sure that they would always be limited as they were in Apollo 10 and hence that they would not lead to a fuel-cell failure.

Our investigation revealed that small, isolated disturbances in fuel-cell temperature were often present, as figure 1-6 shows. Pratt & Whitney, North American, and NASA then performed a detailed stability analysis of the fuel-cell system, transfer functions were experimentally determined, and finally a complete fuel cell test was run to verify the results of the analysis. This work demonstrated that small, isolated disturbances could trigger an instability if the power loading ran sufficiently high and the

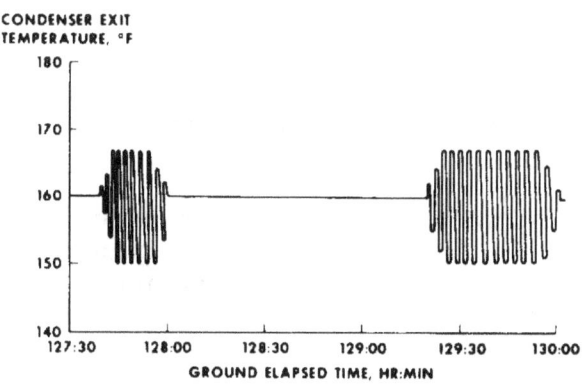

Figure 1-5. - Apollo 10 fuel-cell temperature oscillations as they originally appeared in flight.

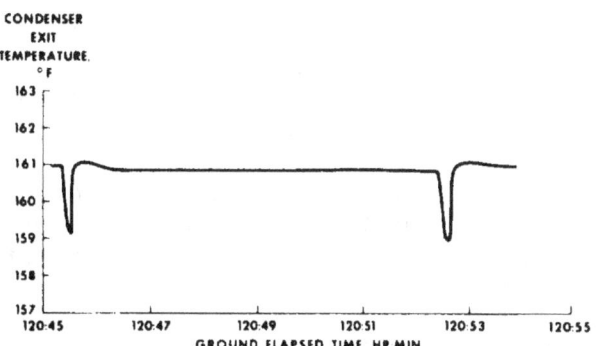

Figure 1-6. - Disturbance of Apollo 10 fuel-cell temperature as it was identified in the laboratory.

temperature sufficiently low. The analysis also showed that the amplitude of the oscillations would always be limited as it was in Apollo 10. With this information, it was possible to devise procedures to eliminate the oscillations, should they occur.

The solution as described here probably sounds simple. Yet, a similar task, formulated as a research assignment, might have taken a year or more to complete. Here, closeout of the failure was done in weeks.

The fuel-cell anomaly was only one example of a discrepancy. The total task — handling all flight anomalies — was enormous; yet, it was completed before each flight.

FLIGHT MISSIONS

It is difficult to describe, to those not directly involved in the Apollo Program, just how much work went into operational activities. First, we had to decide the kinds of mission to be flown: What would be the best series of missions to achieve a successful manned lunar landing at the earliest time? Then these missions had to be planned in detail: How should each mission be designed to meet the largest number of operational and hardware objectives, even in the event of unplanned events? (Operational objectives are concerned with guidance, navigation, trajectory control, rendezvous, etc.; hardware objectives are concerned with the verification of each system or subsystem in the flight environment.) Finally, plans had to be made for the execution of the mission: Detailed rules were evolved for every imaginable contingency; the proper flight-control displays were defined to permit instantaneous reaction to emergencies, and countless hours were spent in simulations of every conceivable situation.

Mission Definition

Early in 1967 the situation was as follows. Many development flights had taken place to test the launch-escape system under extreme conditions, to test the command module heat-protection system at speeds halfway between earth-orbital and lunar-reentry velocities, and to put the guidance and propulsion systems through their preliminary paces. However, Saturn V had not yet been flown, reentry at lunar-return speeds had not yet been made, the lunar module had not yet been flown, and man had not yet been in space in Apollo hardware.

The flight-test program shown in figure 1-7 was then evolved through an iterative and flexible process that was changed as time went on to take the best advantage of knowledge about mission operations and hardware availability at any given time. The basic principle in planning these flights was to gain the maximum new experience (toward the goal of a lunar landing) on each flight without stretching either the equipment or the people beyond their ability to absorb the next step.

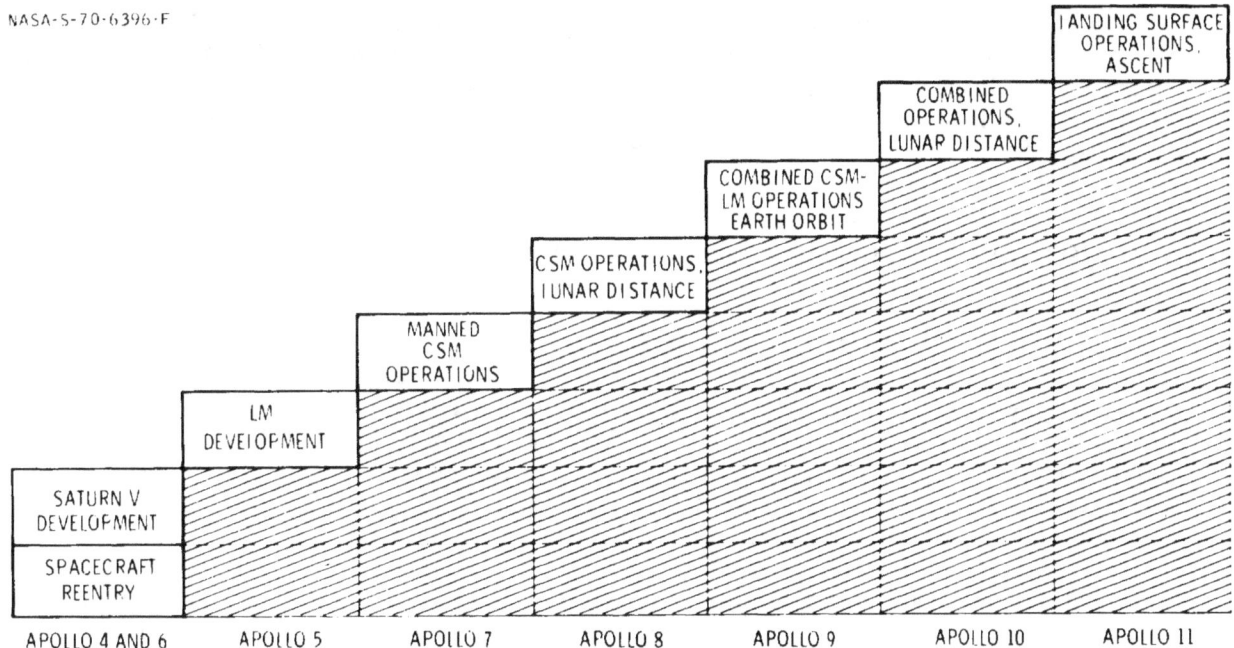

Figure 1-7. - Buildup of Apollo mission capability.

Too small a step would have involved the risk that is always inherent in manned flight, without any significant gain — without any real progress toward the lunar landing. Too large a step, on the other hand, might have stretched the system beyond the capability and to the point where risks would have become excessive because the new requirements in flight operations were more than people could learn and practice and perfect in available time.

Apollo 4 and 6 saw the first flights of Saturn V. Apollo 4 was almost letter perfect. Yet a repeat flight was planned and, in retrospect, proved to be very important. Serious defects in the Saturn propulsion system and in the spacecraft adapter, that were not apparent on Apollo 4, caused major failures on Apollo 6. These failures led to an extensive ground-test program and to hardware changes before the next flight of the launch vehicle. Apollo 4 also served to qualify the spacecraft heat shield under severe simulated lunar reentry conditions; the flight showed that the design was conservative.

Apollo 5 was an unmanned flight of the lunar module. The lunar module guidance system, both propulsion systems, and the all-important staging sequence between the ascent and descent stages functioned well.

The first manned flight of the command and service module came in October 1968, with Apollo 7. The spacecraft performed beyond all expectations in this 11-day flight. Each of the command and service module systems (except the docking system) was put through its paces without a significant malfunction.

The decision to fly into lunar orbit on Apollo 8 came relatively late. It was made, on a tentative basis, in August 1968. At that time, the test experience with the command and service module had been very good. The lunar module schedule, on the other hand, was slipping; and the first manned lunar module exhibited the normal "first ship" difficulties during checkout at the NASA Kennedy Space Center. Also, a detailed analysis of results from the unmanned Lunar Orbiter Program had shown that navigation around the moon would present many unexpected computational difficulties.

For all of these reasons, it was decided that Apollo 8 should be a CSM-alone lunar-orbit flight. This decision was reaffirmed with the success of Apollo 7, and the die was cast for making man's first flight to the moon in December 1968.

In Apollo 9, both spacecraft, the lunar module and the command and service module, were tested together for the first time. First, all of the lunar module systems were tested in manned flight. Then methods for the following spacecraft operations were worked out and verified: communications between two spacecraft and the ground; tracking, guidance, and navigation; and rendezvous and docking. Also on Apollo 9, the extravehicular mobility unit (the lunar space suit and its life-support system) was tested in the actual space environment.

After Apollo 9 another decision had to be made: Were we then ready for a lunar landing, or was the step too big? We decided that we faced too many remaining unknowns: performance of the lunar module in the deep-space environment, communications with the lunar module at lunar distances, combined operations with two spacecraft around the moon, rendezvous around the moon, and, of course, the lunar descent, landing, surface operations, and ascent. In lieu of a landing, we planned to do as many of these tasks as possible on Apollo 10 without actually touching down on the surface of the moon.

The entire series of flights represented a step-by-step buildup, with each step leading closer to a lunar-landing ability. Our intent was to use the procedures developed on one flight on each subsequent mission. Changes were allowed only if they were essential for flight safety or mission success. By means of this buildup, we minimized the remaining tasks (descent, landing, surface operations, and ascent) that could be worked out only on the actual landing mission. The Apollo 11 crew was able to concentrate on these remaining tasks, to work them out in detail, and to carry them out with perfection.

Mission Planning and Execution

Once basic missions had been defined, each flight had to be planned in detail. The mission planner tries to fit into each flight the maximum number of tests of the hardware and the widest variety of operations. For example, he will develop a rendezvous profile for a single earth-orbital flight that involves all of the normal and abnormal rendezvous conditions which might be encountered around the moon — rendezvous from above, rendezvous from below, rendezvous with the lunar module active, rendezvous with the command and service module active, and rendezvous with varying lighting conditions. At the same time, the mission planner will try to exercise all of the propulsion systems and all of the navigation systems on both spacecraft.

After mission plans come the mission techniques (by another name, data priority). Given two or three data sources (for trajectory control), which of the sources should be believed and which discarded? Limits for each system had to be determined, and logic flows for every conceivable situation had to be developed.

Finally, the flight controllers take over. They had participated, of course, in the mission-planning and mission-technique activities; but now they had to work out each step of the flight and anticipate every emergency situation that might arise. What is the proper action when one fuel cell fails? What if two fail? The answers to thousands of questions like these had to be derived in terms of the specific mission phase. A rendezvous radar failure before command and service module-lunar module separation dictates that the two vehicles should not be separated. The same failure after separation allows the mission to be continued because the risk of rendezvous without radar has already been incurred and will not increase in subsequent mission sequences. Each of these events was documented as a mission rule long before the flight, and mission rules were placed under "configuration control," as was every other aspect of the Apollo system.

Flight controllers also worked out the best formats for their real-time displays. During the Apollo 11 descent to the surface of the moon, the flight controllers could watch, with a delay of only 6 to 10 seconds, the functioning of nearly every onboard system. They saw the rise in chamber pressure as the descent engine was throttled up to full thrust, and they could determine that the throttle-down occurred at the proper time. The flight controllers could also compare the descent trajectory from three data sources — two onboard guidance systems and the ground tracking system. With this information, a flight controller on the ground could tell the crew, nearly 250 000 miles away, to ignore the alarms from the onboard computer during the most critical portion of the descent, because the system was guiding the spacecraft correctly.

Many of the techniques used during the flight were developed during countless hours of simulations. Simulation is a game of "what-if's." What if the computer fails? What if the engine does not ignite? What if ... ? The game is played over and over again. The flight controllers do not know what situation they will face on the next simulation. By the time of flight, they will have done simulations so often and they will have worked together as a team so long, that they can cope with any situation that arises.

Because the Apollo equipment has worked so well and because there have been so few contingency situations, one could conclude that much of the planning, many of the mission techniques, and much of the training were done in vain. But this is an incorrect conclusion. As a minimum, the state of readiness that evolved from these efforts gave us the courage and the confidence to press on from one mission to the next. Also, there were situations — the computer alarms during the descent of Apollo 11 and the lightning discharge during the launch of Apollo 12 — that might have led to an abort if the team had been less well prepared and less ready to cope with the unexpected.

FLIGHT CREW TRAINING

The first six Apollo manned flights carried 18 astronauts — all professional pilots, skilled and superbly trained. Altogether, they had flown on 18 flights in Mercury and Gemini before they flew in an Apollo spacecraft. Five had flown twice before, eight had flown once before, and five flew for the first time in Apollo.

Training for Apollo is not easy. Two highly sophisticated machines are involved, each far more complex than those in Gemini. The astronauts had to become expert in the workings of both spacecraft. They became computer programmers and computer operators, space navigators, guidance experts, propulsion engineers, fuel-cell-power managers, environmental-control-system experts — to mention but a few areas of expertise. Of course, they had to learn how to control and fly two spacecraft with vastly different handling qualities under conditions of launch, translunar flight, lunar-orbit flight, lunar landing, lunar launch, rendezvous, docking, transearth flight, and reentry.

The astronauts used a variety of training devices — high-performance airplanes to stay alert and sharp; a special dynamic launch simulator to practice manual takeover and abort modes; mission simulators to duplicate here on earth every spacecraft function and display under all possible conditions; partial-gravity simulations under water, in airplanes, and on a special servo-controlled device on the ground; a docking trainer; and a flying lunar-landing training vehicle that has a jet engine to take out five-sixths the gravity of earth so that the vehicle has the same flying characteristics as the lunar module has on the moon.

The astronauts also needed plans and procedures. Flight plans spelled out each step of the mission. Detailed "time lines" were developed for every function that had to be performed, minute by minute. Crew procedures and checklists were an adjunct to the flight plan. The step-by-step sequence for each spacecraft activity, each maneuver, each propulsive burn was worked out well in advance and was used again and again during practice and simulation.

Configuration control was as important in the astronaut training as in every other category. Simulators had to look just like the spacecraft to be useful, and last-minute spacecraft changes had to be incorporated in the simulators as well. Crew procedures that had worked well on one flight could not be changed, through "crew preference," for the following flight.

Pete Conrad said that landing his Apollo 12 lunar module, after dust obscured the landing point, was the most difficult task he had ever performed. It took all of his 20 years of experience as a professional aviator, his previous work on two Gemini flights, his training for Apollo, and his knowledge and confidence in the Apollo spacecraft systems to make that landing a success.

CONCLUDING REMARKS

Spacecraft development, mission operations, and flight crew activities — in reviewing these areas of Apollo, I see one overriding consideration that stands out above all the others: Attention to detail. Painstaking attention to detail, coupled with a dedication to get the job done well, by all people, at all levels, on every element of Apollo led to the success of what must be one of the greatest engineering achievements of all time — man's first landing on the moon. The reports which follow amplify this observation.

2. DESIGN PRINCIPLES STRESSING SIMPLICITY

By Kenneth S. Kleinknecht
Manned Spacecraft Center

The shaping of reliable, safe Apollo spacecraft owes its success to specific principles stressing the simplicity, both in originating and evaluating hardware designs. The primary consideration governing the design of the Apollo system was that, if it could be made so, no single failure should cause the loss of any crewmember, prevent the successful continuation of the mission, or, in the event of a second failure in the same area, prevent a successful abort of the mission.

DESIGN PRINCIPLES

To implement this policy, the following specific principles were established to guide Apollo engineers when they originated and evaluated hardware designs.

1. Use established technology.

2. Stress hardware reliability.

3. Comply with safety standards.

4. Minimize inflight maintenance and testing for failure isolation, and rely instead on assistance from the ground.

5. Simplify operations.

6. Minimize interfaces.

7. Make maximum use of experience gained from previous manned-space-flight programs.

Established technology was used for areas in which performance and reliability goals had already been realized. Hardware design precluded, as much as possible, the necessity to develop new components or techniques. When this policy could not be met, procedures were established whereby management approved new development requirements only after clear-cut plans for the development effort and a suitable backup capability had been defined.

A primary criterion governing a particular system was whether or not the design could achieve mission success without incurring risk of life or serious injury to the crew. Numerical values for reliability standards and minimum mission objectives were established. Trade-off studies of design and performance were then made to define the necessary redundancy (including alternate or backup equipment as well as modes of operation) for meeting mission goals within the program constraints of time, cost, and weight. Apollo engineers performed not only comprehensive failure mode and effects analyses, but also single-point failure analyses. Through a series of iterative design reviews, the engineers eliminated or minimized each potential failure point.

Safety considerations were emphasized by selecting appropriate design features and proven, qualified components and operating principles. Integrated safety analyses defined the interfaces between subsystems. Thus, safety problem areas were identified for the combined system. Failure modes considered included structural failures, ruptures, fuel leaks, hose-tubing failures, electrical open-short, and fastener failures.

During Apollo spacecraft design and planning, inflight maintenance was carefully considered, but the disadvantages of this approach far outweighed the advantages. In consideration of the duration of the Apollo missions, reliable performance could be achieved through component, circuit, and system redundancy, since the subsystems are not required to operate over long periods of space travel or after being dormant for long periods. The additional connectors and test points required for an inflight maintenance also significantly degraded the overall reliability of the system. Of lesser consideration was the provisioning and stowage of the necessary spare parts within an already limited volume.

With elimination of maintenance and failure isolation by the crew in flight, data were made available to allow the ground to troubleshoot, isolate the failure, and recommend corrective action. This reliance on the ground for troubleshooting has proved quite effective, since subsystem operational specialists (flight controllers) and design specialists (subsystem engineers) are available and free to concentrate continuously on the resolution of spacecraft problems. This practice also relieves the crew from a training requirement of becoming intimately familiar with the detailed subsystem design. In addition to having cockpit display information relayed from the crew, the ground has approximately 330 data channels through telemetry and approximately 1100 ground or preflight checkout data channels.

Rapid data storage, comparison, retrieval, and analysis by computer complexes within the Mission Control Center at the NASA Manned Spacecraft Center (fig. 2-1) give the ground an enormous advantage over the crew, which must continue to operate the spacecraft, eat, and sleep on a fixed schedule. In addition, the ground has complete files of drawings and specifications available. The ground also has simulators — exact functional duplicates of the flight spacecraft — to evaluate flight problems and corrective procedures.

The Apollo maintenance concept, although not providing for inflight maintenance, does permit removal and replacement of "black boxes" during preflight checkout. This procedure does not require entering the box interior, a practice which could disturb adjacent or related assemblies. Performance of the replaced equipment must be at least equal to the required performance of the original system. To isolate faults at the black-box level, test points are located in the subsystems. The engineers have appropriate ground-checkout equipment for fault isolation.

Guided by concepts of simplified functional operation, Apollo engineers combined off-the-shelf components into integrated systems which performed so efficiently that the crew was permitted to devote the majority of its time to the productive tasks of scientific experimentation and data acquisition. Some design requirements, however, resulted not so much in simple mechanisms as in extreme simplicity and reliability of operation. Thus, one crewman, wearing a pressurized space suit, can perform all critical spacecraft control functions.

To achieve a minimum of interfaces, subsystem designs were developed and tested independently and later joined with other spacecraft subsystems. The final Apollo configuration was the result of technological and weight constraints. The Apollo external interfaces between the launch complex and the launch vehicle and the internal interface between the command module and the lunar module are defined in detail by interface control documents, and have been carefully screened to eliminate all but essential functions, thus keeping vehicle interfaces to the minimum. For example, there are only 36 wires between the lunar module and command and service module and only some 100 wires between the spacecraft and the launch vehicle.

To use the experience gained from Project Mercury and the Gemini Program, engineers with operational background from these programs were involved in all major

Apollo design reviews. This procedure allowed incorporation of their knowledge as the Apollo design evolved. This involvement proved a key factor in producing spacecraft that have performed superbly so far. Even the Apollo 13 oxygen tank rupture, by far the most critical problem of any Apollo mission to date, was overcome by relying on preplanned emergency procedures and the resourcefulness and ingenuity of the astronauts and the ground support team.

DESIGN FEATURES

Apollo gains a measure of simplicity from features simple both in design and operation, complex in design but simple to operate, or simple by being passive in function. The concept of simple design and simple operation is best illustrated by the Apollo rocket-propulsion systems (fig. 2-2). The pressure feeding and redundant valving guarantee the arrival of the propellant in the combustion chamber, where hypergolic reaction assures ignition. Ablative materials for chamber walls assure chamber integrity while simplifying design greatly.

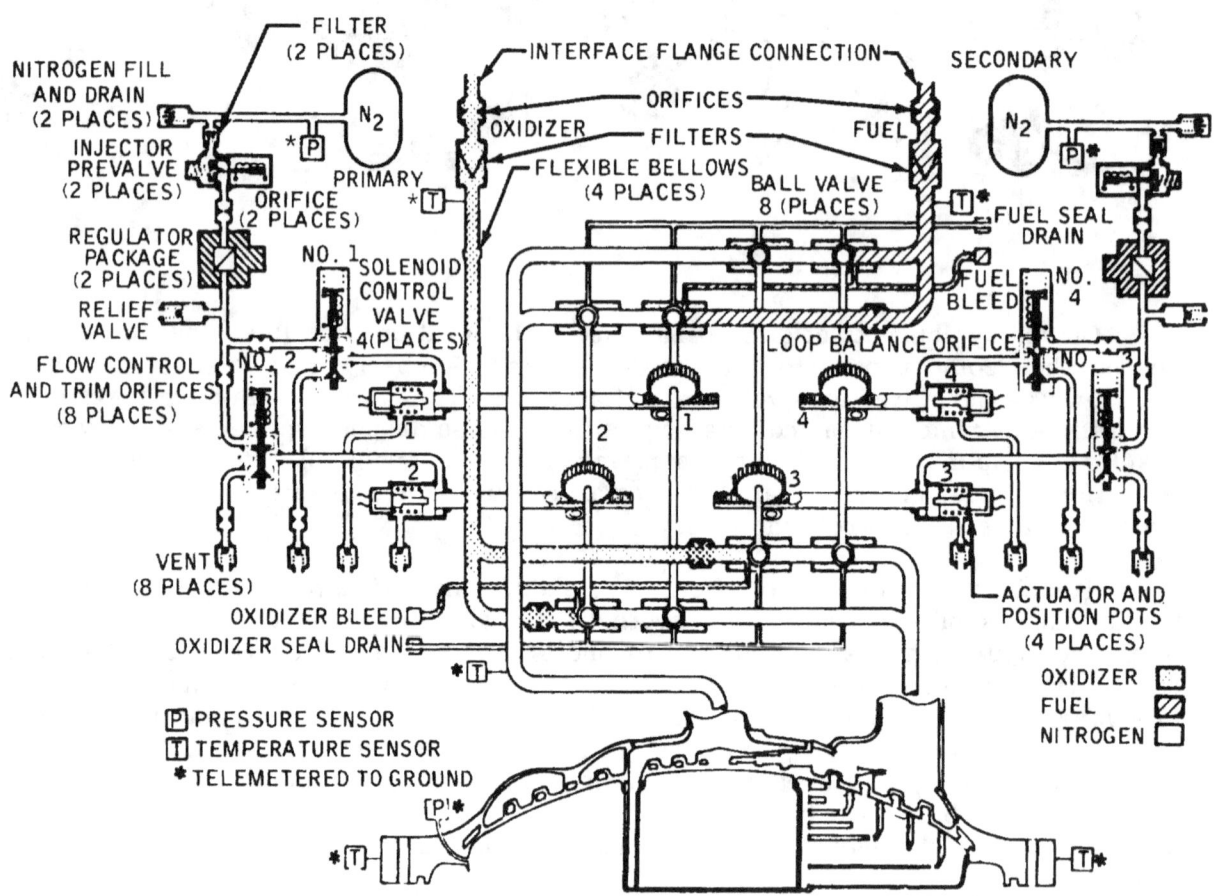

Figure 2-2. - Service propulsion engine propellant valve and injector.

The latching device for the crew hatch, (fig. 2-3) illustrates a complex but simply operated mechanism. Although the device contains approximately 400 parts, it allows a crewman, with a single movement of his arm, to open the command module hatch in less than 10 seconds.

Some design features are simple by being passive (for instance, thermal control). Thermal coatings, ablative heat shields, and insulation eliminate the electrical power requirements of an active system and necessitate only attitude adjustment to maintain spacecraft temperatures within acceptable tolerances.

Apollo reaction-control systems, in both the lunar module and the command module, represent prime examples of redundancy. The command and lunar modules have two parallel and independent systems, either of which is able to meet mission requirements.

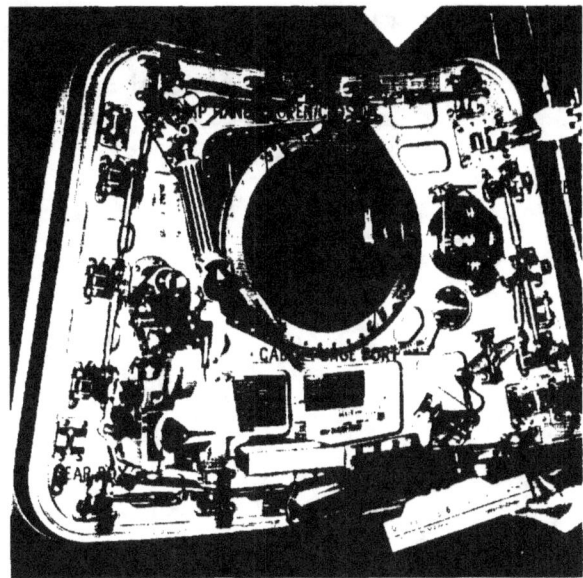

Figure 2-3. - Command module hatch.

Critical events initiated by pyrotechnic devices and the cooling of temperature-sensitive subsystems by the environmental control system (ECS) represent two examples of redundant paths. In the case of pyrotechnic devices, two separate wire runs and initiators receive the same event signal. Likewise, the ECS contains two water/glycol circulating plumbing loops, each having its own control system. Not all systems serviced by the primary glycol loop can be supported by the secondary system, but enough capability exists to return to earth safely.

For some of the critical systems, redundancy is not provided by duplication. For example, the lunar module abort guidance system provides virtually the same ability for delivering the lunar module back into orbit from the descent trajectory and from the lunar surface that the primary guidance system does. However, the designs of the abort and primary guidance systems, both hardware and software, are completely different. In at least one respect, this difference improves reliability by eliminating the possibility of common design faults, particularly in the computer programs, although this was not the basic reason for using this approach. In addition, some simultaneous component failures of both systems can even be tolerated by adopting a more manual mode of operation and using the remaining capability of each system.

CONCLUDING REMARKS

The Apollo design philosophy has resulted in a highly reliable spacecraft capable of placing man on the moon and returning him to earth safely. Simple design practice, coupled with stringent technical and administrative discipline, has achieved this end.

The spacecraft (the command and service module and the lunar module), embodying millions of functional parts, miles of wiring, and thousands of welded joints, has evolved into a truly operational space transportation system.

3. TESTING TO ENSURE MISSION SUCCESS

By Scott H. Simpkinson
Manned Spacecraft Center

All programs, once the overall objectives are fixed, must meet well-defined design goals, or management fails. However, this hard fact does not prevail in the world of test. Theoretically, one no longer needs to test hardware or software, after developing a new concept, except occasionally to gather empirical data needed to operate the equipment. Unfortunately, designs are not perfect, materials and processes often do not work as the designers expect, manufacturing techniques sometimes inadvertently alter the design, assembly procedures leave room for mistakes, engineering and development tests do not necessarily provide all the required data, and, finally, substandard workmanship and human error creep in.

All of these factors require attention at the outset of a program. Some factors, such as human error, demand vigilance until delivery of the last item. Experience has shown that only a well-balanced test program can instill confidence in the delivered hardware and software for a space vehicle.

At the beginning of the Apollo Program, high priority went to setting up a program for one-time qualification of a component or system design and to manufacturer execution of the program. All contracts contained specific clauses relating to qualification tests. These tests provided a reasonable margin of safety, taking into account the expected environments the pieces would pass through during storage, transportation and handling, ground-test duty cycles, and two-mission duty cycles. After the early unmanned flight-test program had started, actual measurements of the launch environment led to adjustments in the qualification vibration levels. Equipment already tested at too low an amplitude had to pass an additional (or delta) qualification test program. Rigorous monitoring and careful failure reporting allowed correction of design and process failures.

Even with this exacting qualification program, a number of experienced engineers believed each flight item should have to pass some environmental testing before NASA accepted it for installation and flight. Thus, nearly all functional components and systems underwent acceptance testing. However, the detailed test plans were left in the hands of the individual designers and system engineers. During the early stages of the Apollo Program, most components and systems were limited to a complete functional bench test at room temperature and pressure and a survival test after a brief exposure to random vibration in the axis suspected of being the most sensitive and at the expected flight-power spectral density.

A few vendors, who were experienced on other critical military and NASA programs and who were supplying electronic components for the lunar module (LM), also performed temperature-limit tests at their own discretion during buildup or during the final acceptance tests (or, in some cases, during both). Unfortunately, the expected flight vibration levels were so low in many cases that these early environmental acceptance tests did not reveal errors of workmanship and manufacturing processes. Many system manufacturing and processing errors came to light late in the cycle, delaying the program and wasting manpower.

After the spacecraft fire, NASA launched an extensive review of the Apollo acceptance test practices. Subcontractors and vendors for 33 Apollo spacecraft assemblies, representing a cross section of electrical, electronic, and electromechanical equipment throughout the spacecraft, received 79 detailed questions concerning their individual acceptance test plans and objectives. This survey revealed the inadequacy of environmental acceptance tests or, in many cases, the nonexistence of acceptance tests. Soon, both the command and service module (CSM) and the LM would carry men for the first time. The decision was made by NASA to review completely all Apollo spacecraft acceptance, checkout, and prelaunch test plans and procedures.

In general, the review found factory checkout and prelaunch tests at the launch site adequate and, in many cases, with overly tight tolerances. Between installation and launch, the equipment passed the same tests a number of times.

The revised overall testing ground rules, which came out of the review, resulted in a more efficient test plan from predelivery acceptance tests to launch. However, the results of the environmental acceptance test review carry much more significance for those who will make the decisions for future programs.

The prime contractor for the LM required nearly all subcontractors and vendors to subject their equipment to 1 minute of random vibration in each of three mutually perpendicular axes. However, most of the vibration levels were very low, as table 3-I shows. A decision early in the Apollo Program had set acceptance vibration test levels 6 decibels below those for qualification, or at 0.005 g^2/Hz, whichever was greater. (On the average random vibration table, one cannot practically set up a vibration level lower than 0.005 g^2/Hz.) Many of the components did not have to function or pass continuity checks during vibration tests, only before and after.

To determine proper acceptance vibration tests, a study reviewed 20 major aerospace programs under nine different prime contractors. Bearing in mind the true purpose of acceptance testing — to prevent the installation and flight of substandard equipment — NASA combined the results with an understanding of the nature of the failures encountered after acceptance testing. The resulting program fashioned Apollo vibration tests after those for Gemini.

A component would have to withstand the vibration levels shown in figure 3-1 in each of three mutually perpendicular axes for a minimum of 1 minute and a maximum of 5 minutes. In addition, a firm ground rule pegged the minimum qualification vibration test level at 1.66 times the acceptance test level at all frequencies. In addition, testers had to monitor all pilot-safety functions and continuously check all electrical paths for continuity and short circuits during each of the three vibration cycles. The testers also had to monitor all mission-success functions, if at all feasible, within schedule and cost constraints. Of the original acceptance vibration test plans for approximately 150 deliverable LM items, 80 plans were changed significantly.

TABLE 3-I.- LUNAR MODULE ACCEPTANCE VIBRATION TEST IN 1967

Subsystem	Quantity to be vibrated	Not vibrated	Number vibrated at vibration level, g^2/Hz				
			<0.01	≥0.01 <0.02	≥0.02 <0.03	≥0.03 <0.04	≥0.04
Communication and data	35		6	6	2		21
Electrical power	19		10	3	1	1	4
Environmental control	6		5				1
Guidance, navigation, and control	16		4		7	1	4
Propulsion	7	1	4				2
Reaction control	2			1			1
Crew provisions	5		1	2		1	1
Displays and controls	22		4	12	2	2	2

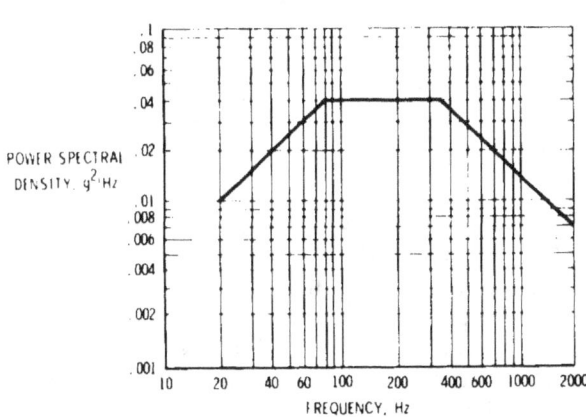

Figure 3-1.- Revised Apollo acceptance vibration test guidelines.

A special procedure governed mission-success equipment that could not be removed from LM 3 and LM 4 without a serious impact. Similar equipment for later spacecraft underwent the new tests. Some of the test failures indicated the possibility of similar failures on LM 3 and LM 4. Depending on the failure probability and the impact of the failure, we either changed to newly tested equipment or accepted the risk.

A similar review of the CSM again found vibration levels too low to detect workmanship errors. A considerable number of the components experienced only single-axis sine-wave excitation. Of over 200 deliverable items of CSM equipment tabulated in table 3-II, the requirements for 80 were changed.

TABLE 3-II.- COMMAND AND SERVICE MODULE ACCEPTANCE VIBRATION TEST IN 1967

Subsystem	Quantity to be vibrated	Not vibrated	Number vibrated at vibration level, g^2/Hz				
			< 0.01	≥0.01 < 0.02	≥0.02 < 0.03	≥0.03 < 0.04	≥0.04
Communication and data	33		11	3	1	16	2
Electrical power	24	10	3				11
Environmental control	65	64		1			
Guidance, navigation, and control	28		10	3	1	3	11
Propulsion	11	6			1	3	1
Reaction control	1						1
Sequential events control	7		1	2	2		2

The need for thermal or thermal vacuum testing as a tool for finding workmanship faults became apparent during the review of components for reacceptance vibration testing. The construction of certain items prevented vibration tests from revealing critical workmanship errors. As a result, some items were deleted from the acceptance vibration test list and required an acceptance thermal test. The review turned up several Apollo components subject to acceptance thermal tests for this reason. However, the review team found the criteria governing acceptance and qualification thermal testing and the relation between the two to be unacceptable.

A joint meeting of North American Rockwell, Grumman Aerospace, Massachusetts Institute of Technology, Boeing, and Manned Spacecraft Center representatives replaced the old standards shown in table 3-III with the new standards graphed in figure 3-2. The new guidelines called for 1-1/2 temperature cycles with a swing of 100° F or more, starting and ending at room temperature. The guidelines specified holding the test article at the two high-temperature and one low-temperature limits for 1 hour after the temperature had stabilized. The equipment should operate throughout the test and undergo continuous monitoring for continuity. It should pass complete functional tests immediately before and after the thermal test, and an adequate functional test after stabilization at the high and low temperatures. Equipment suspected of being adversely affected by temperature gradients should also complete functional tests during the two transitions between the high-temperature and low-temperature limits. An arbitrary decision set the acceptance test limits at 20° F less than the

TABLE 3-III.- CRITERIA FOR LUNAR MODULE QUALIFICATION

AND ACCEPTANCE THERMAL TESTING IN 1967

Parameter	Cold-plate cooled	Radiation cooled
Qualification		
Pressure, torr	1×10^{-5}	1×10^{-5}
Temperature, °F:		
Root of flange	35 to 135	Not controlled
Environment[a]	0 to 160	0 to 160
Acceptance		
Pressure, torr	1×10^{-5}	Ambient
Temperature, °F:		
Root of flange	35 to 135	Not controlled
Environment[b]	0 to 160	30 to 130

[a]Equipment thermally isolated for 24 hours at each level.

[b]Equipment thermally isolated for 4 hours at each level.

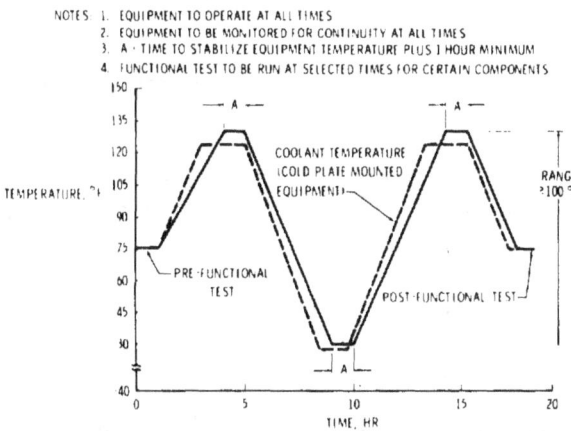

Figure 3-2.- Revised Apollo acceptance thermal test guidelines.

qualification test limits. Equipment normally cooled on a cold plate should be mounted on one during the test with the coolant entering the cold plate externally controlled to between 10° and 15° F cooler than the environment. Of approximately 260 LM and 215 CSM items of deliverable equipment reviewed, 70 LM and 55 CSM items required additional or new acceptance thermal tests to augment or take the place of the acceptance vibration tests.

By December 1969, over 15 000 tests had been performed to the revised environmental acceptance test requirements. The results are shown in figures 3-3 and 3-4, and installation of acceptance-tested crew equipment for the command module is shown

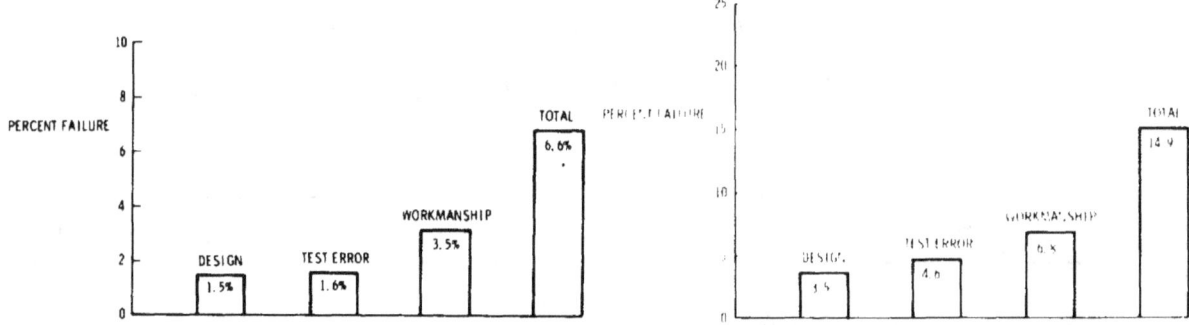

Figure 3-3. - Results of acceptance vibration tests for 11 447 tests of 166 different components.

Figure 3-4. - Results of acceptance thermal tests for 3685 tests of 127 different components.

in figure 3-5. While workmanship errors accounted for the majority of failures, design deficiencies not revealed during the qualification tests caused more failures than expected. Because of the design problems and unavoidable test errors, the startup phase was painful. Individual failure rates often exceeded 25 percent. Some exceeded 50 percent. With many new components entering environmental acceptance test during the first year of the reacceptance test program, design and test error rates came down slowly. However, both finally reached approximately 1.5 percent and have stayed there for the past 9 months. Workmanship errors, on the other hand, have remained relatively constant at approximately 5 percent.

In retrospect, several recommendations and points of interest stand out from the test experience gained during this nation's three major manned spacecraft programs over the past 10 years.

1. Design and development testing plays an important part in the overall test plan. Perform it as early as possible. Document the results well, and hold the data for future reference. Pay particular attention to what seem minor details, especially for substitute parts and "explained" failures.

2. Perform qualification testing at the highest possible level of assembly practical, within reasonable cost and schedule constraints.

3. Before subjecting qualification test specimens to qualification test levels, acceptance test the specimens by following approved test procedures under strict change control. Include applicable environmental acceptance tests.

4. Qualification test specimens should come from among normal production items manufactured and assembled using final blueprints and processes, under normal quality control, and with production tooling and handling fixtures.

5. Make qualification tests rigorous and complete, yet realistic. A strong tendency exists to qualify equipment to the designer's desires rather than to the actual requirements. Where flight equipment will never leave controlled cleanroom conditions and need operate only in outer space after launch, take care not to fall into the classical

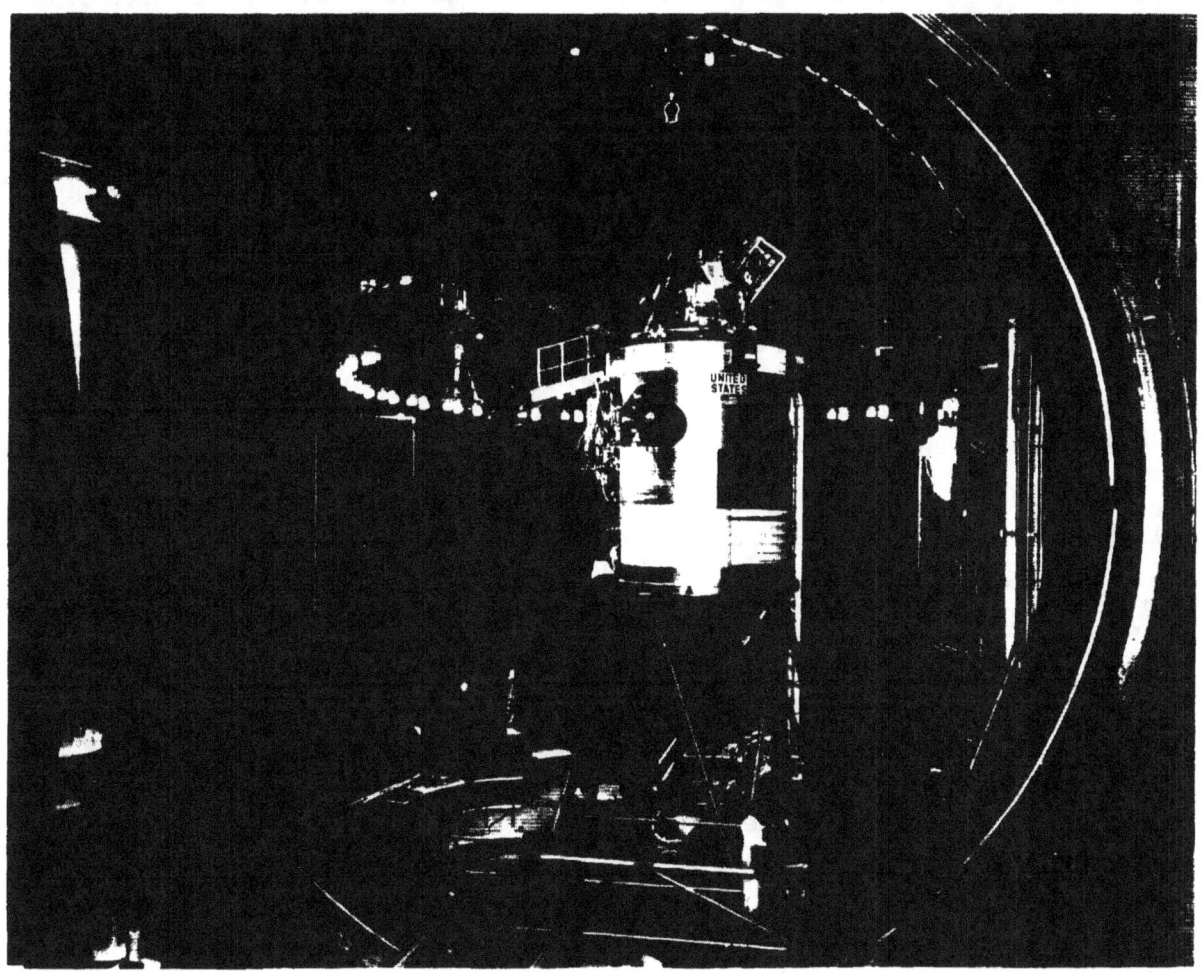

Figure 3-5.- Installation of the acceptance-tested crew equipment in the Apollo command module at the NASA Manned Spacecraft Center.

qualification test programs, which require such things as salt-water immersion, rain and dust, cyclic humidity, fungus, and other environmental extremes.

6. Carefully document, track, explain, and take necessary corrective action on test failures encountered on all production hardware. Qualification test hardware, by definition and from bitter experience, must count as production hardware. No suspected failure encountered during any test on production hardware should escape from this rule, no matter how insignificant or unrelated the failure may seem at the time. Experience has shown that major failures always receive adequate attention. The minor unreported failure is the one that slips by and shows up late in the vehicle test cycle or, worst of all, in flight.

7. Perform qualification vibration testing at excitation levels that provide a 50-percent margin of safety over the expected environments, including acceptance vibration testing. Because of weight and volume considerations, power consumption, or

other limiting factors, this margin may have to decrease. When this becomes evident, treat each case on an individual basis and make sure all parties involved understand.

8. Thermal test for qualification to temperature limits at least 20° F outside the expected temperature limits, including acceptance thermal testing.

9. Monitor the functioning of test specimens during the dynamic phases of qualification and environmental acceptance testing as completely as possible. Test all functions with 100-percent monitoring, including all redundant paths, before and after appropriate phases of the qualification and acceptance tests.

10. Subject to environmental acceptance tests components of critical equipment that must function electrically or mechanically, when the complete deliverable item cannot be visually inspected or functionally tested (or both) for design, manufacturing, assembly, handling, procedural, or workmanship errors.

11. Determine environmental acceptance test requirements on an individual basis, considering what types and what levels of tests reveal quality or workmanship defects. Examine the failure history of each component during the engineering and development stages and during qualification testing. Look at the failure history of like components using similar manufacturing and assembly techniques, particularly those made by the same vendor, in the same plant, and with the same people.

12. Carry out environmental acceptance testing at the lowest practical level of assembly. For example, it is much better to find the solder balls in a sealed relay before building it into an assembly than to cope with an intermittent system failure on the launch pad. The earlier one uncovers a problem and eliminates it from downstream hardware, the less its impact on the overall program.

13. Carefully examine changes of any nature to the hardware for their effect on qualification and acceptance test results. Qualification by similarity can give very misleading results and should take place only with full knowledge of all parties concerned. In the case of acceptance testing, the simple act of removing a cover plate for inspection purposes constitutes potential grounds for a reacceptance test.

14. Total vehicle environmental acceptance tests are desirable. However, tests of this nature become virtually impossible to perform on manned spacecraft. Thorough qualification of spacecraft components, including wiring and tubing installations, combined with proper environmental acceptance tests of the equipment before installation, has thus far assured mission success.

15. Always retest after changes to the hardware or software have been made. Set up rigorous controls to assure it.

16. When possible, test all functions and paths on the installed systems at least once prior to delivery to the launch site. As a general rule, when changes or replacements require retesting, do it at the factory. Prelaunch testing at the launch site should demonstrate total space-vehicle and launch-complex compatibility and readiness. They should not simply prove the adequacy of a given component or single system.

At the start of a program, devise a thorough overall integrated test plan that includes all testing (including engineering and development, qualification, reliability and life, predelivery environmental acceptance, preinstallation acceptance, installed system, altitude, prelaunch, and early unmanned flight tests). The plan should include as much testing as necessary to gain confidence in the hardware, the software, the test equipment, the test procedures, the launch procedures, and the flight crew procedures. The plan should provide for deleting unnecessary phases of testing as confidence grows.

We believe these measures have proved themselves in the Apollo Program. By calculating from the design and workmanship failure rates during reacceptance tests, the program corrected or removed before launch approximately 65 potential spacecraft pilot-safety or mission-critical hardware failures per flight. Some faults remained. Each of the first 10 vehicles flown on the first six manned Apollo missions experienced approximately eight hardware failures. But fewer than two failures per vehicle stemmed from workmanship or quality. Of the total flight failures from these two causes, better or more thorough acceptance testing could conceivably have revealed only five. Also, no evidence in the flight-failure history indicates a failure caused by too much testing.

The real effectiveness of the test program comes out in examining the results of hardware failures during the first six manned Apollo flights. None of the flight failures affected pilot safety or mission success.

4. APOLLO CREW PROCEDURES, SIMULATION, AND FLIGHT PLANNING

By Warren J. North and C. H. Woodling
Manned Spacecraft Center

The formative stages of Project Mercury saw various opinions as to the degree of crew control that should be permitted. To most space planners, Mercury seemed a logical extension of high-performance aircraft flight, in which brief periods of high-altitude zero-g activity had been experienced with no adverse effects. However, there were skeptics who felt that space flight would require radically new procedures and crew operational constraints.

As it developed, Mercury and Gemini attitude control, systems monitoring, and longitudinal translation maneuvers evolved very similar procedurally to aircraft practice. Consequently, the basic concepts for Apollo crew procedures reflect techniques proved in aircraft, Mercury, and Gemini operations.

Over the past decade, space-vehicle designs have incorporated increased astronaut participation to improve vehicle reliability. Mercury used automatically guided military boosters, which were originally designed without consideration for manual monitoring and control. Therefore, it seemed expedient to accept a passive role for the crew. This was done at the expense of developing an elaborate automatic abort-sensing system to protect the pilot in the event of booster failure. The pilot did get a backup abort handle for slow-drift guidance failures that could not be sensed by the automatic system. In Gemini, the complex automatic escape system was eliminated in favor of manual launch-vehicle monitoring procedures and aircraft ejection seats that could be manually triggered if the launch vehicle malfunctioned or if the spacecraft parachute failed during descent.

The enormous TNT equivalency of a "worst case" Saturn booster explosion placed a Mercury-like escape rocket on the Apollo spacecraft. Then concern for a growing control-system failure, coupled with maximum aerodynamic pressure, prompted inclusion of an automatic abort system. Two minutes after lift-off, this automatic system is manually disabled, and the booster firing is terminated manually in the event of a malfunction. Crew displays include the status of launch-vehicle engines and tank pressures. The launch vehicle can be controlled manually if the Saturn inertial platform fails; this ability did not exist in Mercury or Gemini, but was introduced after several years of research and analysis.

Apollo rendezvous, formation flight, and docking maneuvers resemble closely those developed in Gemini. (This, of course, was a major reason for Gemini.) Another significant Gemini carryover is the backup rendezvous procedure, which can be used if the onboard radar, inertial platform, or computer should fail. Placement of rendezvous maneuvers for optimum orbital lighting took advantage of Gemini experience.

Gemini extravehicular activity revealed an anomaly in simulation, which was corrected for Apollo. Short-duration zero-g aircraft parabolas did not reveal the fatiguing effect of long-term tasks in a Gemini pressure suit. For the later Gemini flights, water-immersed simulations (fig. 4-1) provided realistic long-term zero-g workloads. This type of simulation enabled the development of Apollo contingency extravehicular activity (EVA) transfer procedures. An Apollo contingency transfer is required if the lunar module (LM) cannot be docked to the command module (CM) after lunar-orbit rendezvous.

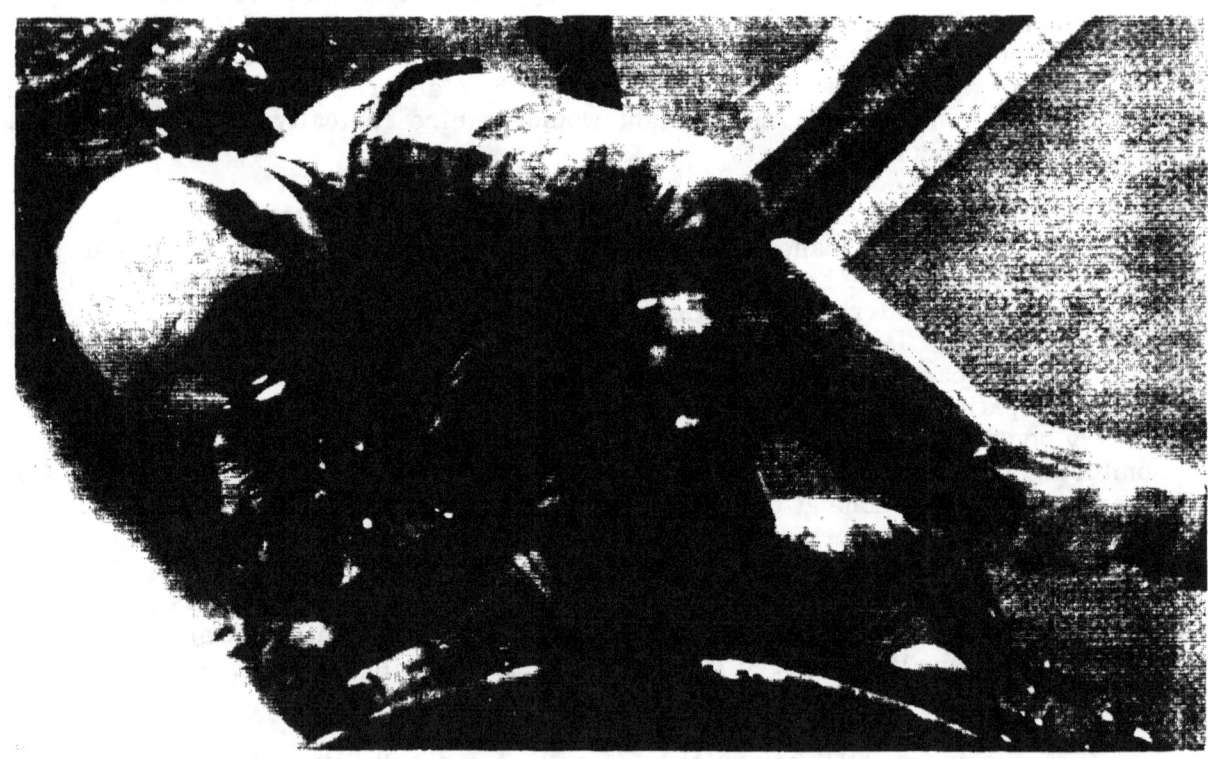

Figure 4-1.- Astronaut trains underwater in simulated zero-g condition in water-immersion facility. Astronaut wears weights on shoulders, wrists, and ankles. Total ballast is about 180 pounds.

Although the spacecraft guidance system normally specifies the proper reentry roll angle that is required to maneuver down range or cross range to the target, Gemini and Apollo employ basically the same backup techniques for using the earth horizon to monitor manually and achieve proper reentry roll angles for ranging.

PROCEDURES APOLLO INTRODUCED

Apollo of course encompassed the lunar landing. Although helicopter experience indicated that low-speed maneuvering and vertical descents were straightforward, the lunar landing involved a much higher approach velocity and descent rate, potential visibility problems, and a 1/6g maneuvering environment. Because of the difficulty in training for manual landings in this environment, there were suggestions that we should accept automatic landings and the attendant probability of abort, if the automatic system targeted the spacecraft to a boulder or crater. The obvious desire to land rather than to abort triggered a major effort to build a free-flight training vehicle (fig. 4-2) which would fly in the atmosphere of the earth and would simulate the 1/6g handling characteristic of the LM. Early flights with this vehicle impressed the pilots with the unusually large pitch and roll angles required to achieve and null translational velocities. The flights also indicated that from altitudes of approximately 500 feet, the vehicle could be maneuvered to avoid obstructions after several familiarization flights.

Lunar-surface EVA likewise had no analogy in previous programs. Improved mobility of the Apollo suit, on the other hand, encouraged rapid development of lunar-surface EVA procedures. Three techniques were used to investigate the effects of 1/6g: The KC-135 aircraft (fig. 4-3), the overhead-supported gimbaled harness (fig. 4-4), and the water-immersion facility. Used in a complementary fashion, these simulation devices yielded results which indicated that 1/6g maneuvering stability was no particular problem; in fact, if one started to fall, it was easy to recover because of the low falling acceleration.

Another significant procedural difference between Gemini and Apollo

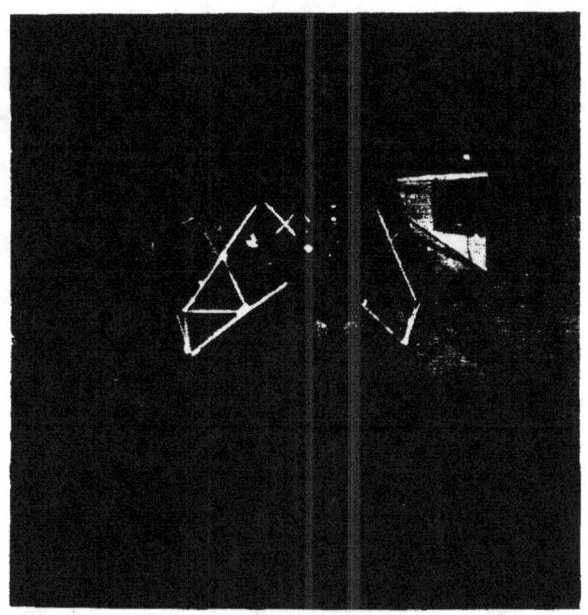

Figure 4-2. - Lunar landing training vehicle trains crews for last 500 feet of altitude in critical moon landing phase.

Figure 4-3. - Lunar module mockup installed in KC-135 aircraft. Support structure takes loads imposed in 2-1/2g pullup, after which zero g is achieved for 20 to 30 seconds on a parabolic flight path.

concerned the ability of the crew, in the event of a ground communications failure, to navigate with the onboard sextant/computer and to calculate spacecraft maneuvers for safe return to the earth. Critical Gemini maneuvers were made within the communications range of ground stations, but the Apollo lunar-orbit retrograde burn and the critical transearth burn occur behind the moon, where crew procedures cannot be monitored by the ground.

NAVIGATION, GUIDANCE AND CONTROL PROCEDURES

Some 40 percent of crew training involves becoming proficient in the use of the three spacecraft navigation systems (one in the CM and two in the LM) and the four guidance systems (two in the CM and two in the LM). The crew interfaces chiefly with these systems through computer keyboards which are about half the size of a typewriter keyboard. Approximately 10 500 individual computer keystrokes are required to complete the lunar mission. The crew must activate individual programs within the computer to execute launch, midcourse navigation, spacecraft propulsion maneuvers, lunar-landmark tracking, lunar descent, lunar platform alinement, ascent, rendezvous, and reentry. Within each program, there are many routines or options. The computer display and keyboard includes three 5-digit registers through which the crew observes displays (such as total velocity, altitude, and altitude rate during the lunar descent).

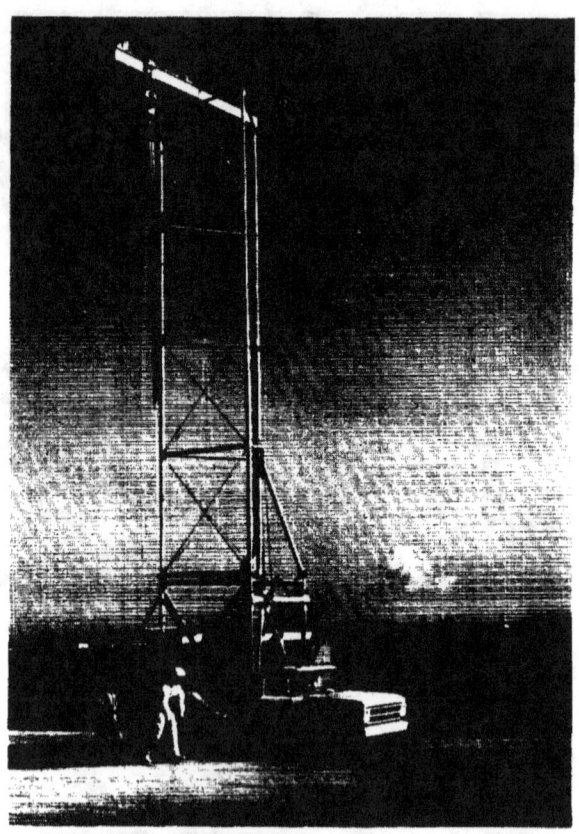

Figure 4-4. - Familiarization run on the mobile partial-gravity simulator used for lunar walk indoctrination.

The primary control systems in both the CM and the LM include digital autopilots which are adjusted through keyboard activity. The crew selects optimum jet configuration, attitude deadband, and maneuver rates. Large variations in vehicle configuration throughout the mission require the crew to be very familiar with control characteristics and handling qualities, which range from sluggish to sporty over the interval from initial booster staging to docking of the lightweight LM.

Much of the crew preflight time goes into reviewing propulsion and guidance malfunction procedures. In this work, alternate system displays are compared to the primary display. If it appears that the primary system has failed, the crew manually makes a switchover. As an example of system redundancy, the initial lunar-orbit insertion maneuver is made with the main service module (SM) engine under automatic control by the primary guidance system, but failure of the primary system can be backed up manually by using the stabilization and control system. If the SM main engine fails during the burn, two crewmembers transfer to the LM, activate the systems, and ignite the descent-stage engine to place the two spacecraft on a return-to-earth trajectory. If the LM descent-stage engine fails, the descent stage is jettisoned, and the LM ascent-stage engine is ignited. The crew controls docked configuration from the LM manually by using the reaction-control system in a translation mode, rather than an efficient attitude-control mode (because of the shifted center of gravity of the docked configuration). A primary guidance system failure of the LM can be backed up by the abort guidance system; as an ultimate control-system backup, either vehicle can be controlled in a manual mode in which the LM pilot control inputs go directly to the reaction jets with no rate-damping feedback.

To provide the crew with this depth and ability to monitor, switch, and control may seem unduly redundant; however, this depth of capability has provided a verification of the adequacy of the primary procedures. Several changes to the primary rendezvous procedures were made, so that the maneuvers could be more easily monitored and remain consistent with the backup procedures.

During descent, the landing site comes into view at the bottom of the LM window at an altitude of 7000 feet, a range of 4 miles, a velocity of 450 fps, and a descent rate of 160 fps. By observing through his window, the commander (in the left-hand crew station) determines if the guidance is targeted to the correct landing area. If not, he can incrementally change the spacecraft flight path 2° laterally and 0.5° up range or down range, by actuations of the right-hand attitude controller.

In the right-hand crew station, the LM pilot is closely monitoring the altitude, the altitude rate, the lateral velocities, and the remaining propellant and is updating the backup guidance system altitude. He calls out altitude and altitude rate to the commander. All of these parameters are compared to nominal and to the limits set as abort criteria. The commander normally takes control in a manual atitude-hold mode at an altitude of approximately 500 feet. At this point, horizontal velocity has dropped to 70 fps and the vertical velocity to 17 fps. The rate of descent is controlled at this time by making discrete inputs to the guidance system at a rate of 1 fps per switch cycle.

Normally some 2 to 4 percent of the descent propellant remains after the landing. Aircraft fuel reserves at landing are considerably greater. Clearly, a critical Apollo crew procedure is propellant monitoring.

The percentage of descent propellant remaining is computed and displayed on digital read-outs. With 2 minutes of propellant remaining, a tank-level sensor is uncovered and a warning light is displayed. The LM descends at a rate of approximately 3 fps. After touchdown and engine shutoff, the crew immediately checks the status of all systems in preparation for an emergency lift-off. If all propulsion, environmental, and guidance systems are satisfactory, preparations are made for powering down the spacecraft and for lunar surface exploration. The subsequent ascent from the lunar surface involves spacecraft power-up, inertial-platform alinement, and a long series of checklist procedures which terminate with descent-stage separation and ascent-engine ignition. During this phase of lunar activity, the two-man crew performs the functions of the entire launch complex and blockhouse crew at Cape Kennedy.

INFLIGHT PHOTOGRAPHY

Besides providing a documentary record of the flight, inflight photography has become a vital part of future mission planning. Television pictures taken several years ago by Orbiter spacecraft formed the base line for constructing the flight maps and the simulator relief model. However, because full coverage of high-resolution pictures does not exist for all of the Apollo landing sites, it has been necessary to obtain additional 70-millimeter still photography of some future sites. (See fig. 4-5.) Coordination of this photography requires careful planning to schedule the photography passes at the proper sun angle and to minimize usage of critical spacecraft attitude-control propellants.

Figure 4-5. - Apollo 12 landing and ascent model of Surveyor and Snowman craters as seen from 1800 feet.

SIMULATION

The development of the Apollo simulation program and associated trainers closely paralleled the development of the Apollo flight hardware and the increasing mission complexity. As flight operations phased from Gemini to Apollo and as the Apollo flights progressed from single to dual spacecraft operations, from earth-orbital to lunar-orbital activities, and finally to the lunar landing, the scope and capability of simulations matured to keep pace with the increasing complexity.

The Gemini Program provided an excellent beginning for Apollo training, because its progress in accurately simulating and adequately training flight crews in the launch, rendezvous, and entry modes was directly applicable. In fact, the first Apollo part-task simulators were converted Gemini simulators. The computer complex and infinity-optics system from the Gemini mission simulators were integrated with simulated crew stations for the CM procedures simulator (CMPS) and the LM procedures simulator shown in figure 4-6. The simulated Agena target vehicle and Gemini spacecraft were replaced with the CM target mockup and the LM crew station on the translation and

Figure 4-6. - Command module procedures simulator and lunar module procedures simulator.

docking simulator, shown in figure 4-7. The dynamics crew procedures simulator, shown in figure 4-8, was converted to the CM crew station for launch and launch-abort training, shown in figure 4-9.

The combined flight of two manned spacecraft during the Apollo 9 mission was a major step in all aspects of flight operations. A corresponding major accomplishment was the addition of the LM mission simulator (LMS), shown in figure 4-10, into the trainer complex and integration of the LMS with the CM mission simulator (CMS), shown in figure 4-11.

The tie-in between the mission simulators and the Mission Control Center at the NASA Manned Spacecraft Center (MSC) marked a significant step in realistically

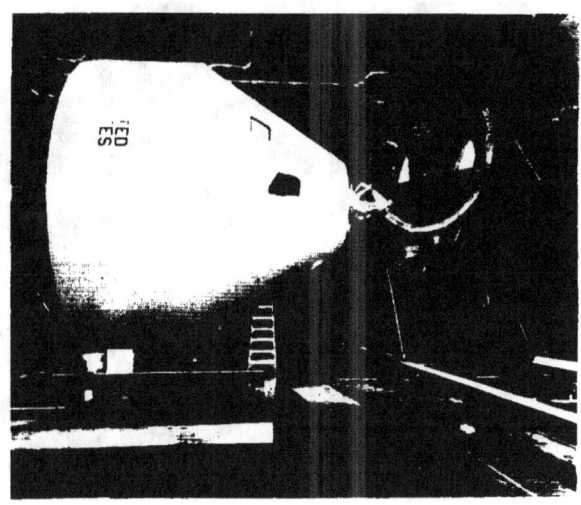

Figure 4-7. - Translation and docking trainer simulates lunar module active docking over last 100 feet of separation distance.

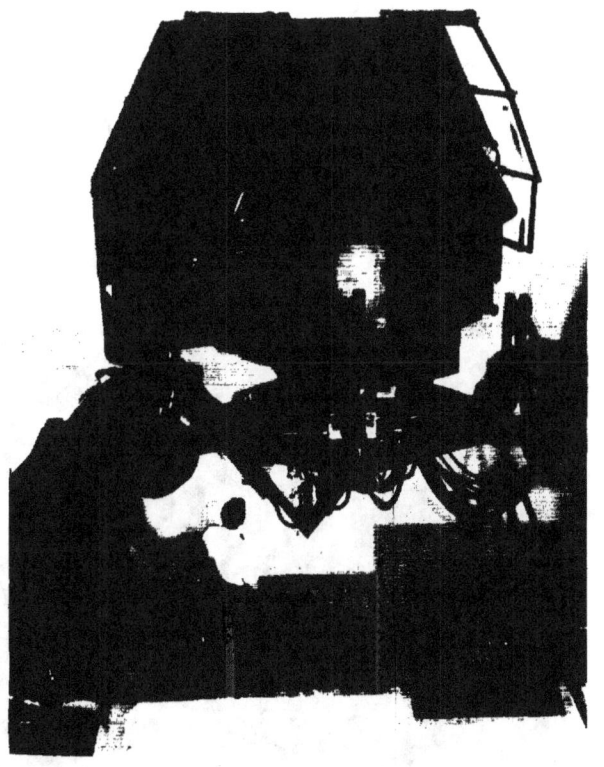

Figure 4-8. - Dynamics crew procedures simulator.

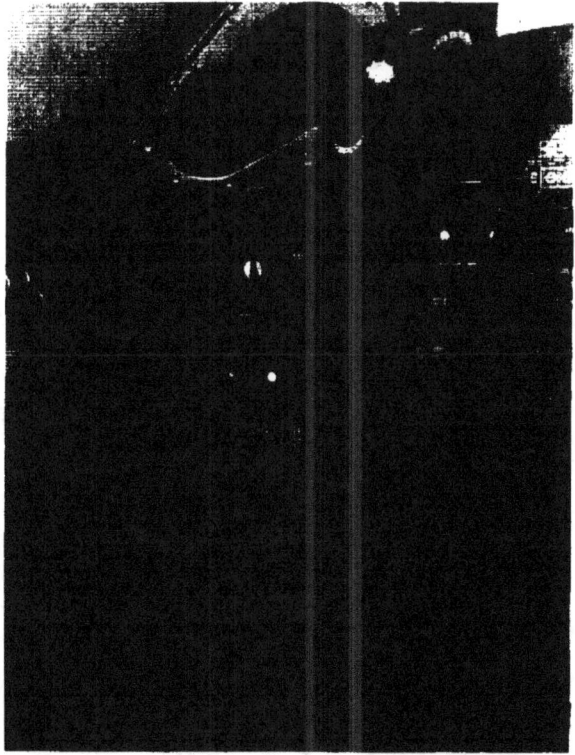

Figure 4-9. - Simulated command module crew station for the dynamics crew procedures simulator.

Figure 4-10. - Lunar module mission simulator with crew station and Farrand optical systems for three windows.

Figure 4-11. - Visual optics and instructor station for command module mission simulator.

conducting space-flight dress rehearsals. The simulations required synchronization of 13 large digital computers, operating together in real time.

The progression of Apollo 8 and 10 to the lunar sphere of influence and the integrated operation of the two spacecraft in lunar orbit required more simulation capabilities, such as additional trajectory equations, new simulator operating modes (fast time, step ahead, safe store) and a complete new set of out-the-window scenes for the crews.

Apollo 11, of course, introduced the final operational phase, landing on the lunar surface. To simulate the lunar landing, it was necessary to build a lunar-terrain model (fig. 4-12) for precise training on the mission simulators of the final-approach and manual-touchdown phases. It was also necessary to use the lunar landing (1/6g) free-flight trainer.

The scope and complexity of the Apollo missions necessitated a comprehensive procedures-verification and crew-familiarization program through a variety of part-task and mission simulations. As mentioned previously, some of the Gemini procedures directly carried over to Apollo, but numerous new areas required further analysis and crew verification. Launch-abort monitoring and launch-vehicle control (Saturn V launch vehicle)

Figure 4-12. - Apollo 12 landing and ascent visual simulation system.

were analyzed in detail at the NASA Ames Research Center and by contractor personnel in separate studies before the backup control of the launch vehicle from the spacecraft was developed. The first flight incorporating this backup mode was on AS-505 with the Apollo 10 flight. The ability to provide pilot control from the spacecraft for the Saturn IVB stage during the translunar injection maneuvers was thoroughly checked out by engineers and pilots on the DCPS and CMS before the Apollo 10 mission.

Considerable revisions were made in the CM computer flight program after the Apollo 7 flight to improve the guidance and navigation logic during rendezvous. Some 1200 man-hours and 100 machine-hours were spent by the MSC engineers with the CMPS in working out the optimum crew displays, procedures, and tracking schedules for the new flight software. Many more hours were spent with the flight crews on the CMPS and CMS familiarizing and training them for the Apollo 9 mission.

In lunar descent and landing, simulation and training emphasized the monitoring of the primary guidance system and the takeover by the commander for the final phase of manual control to touchdown. More than 220 landings were flown on the LMS by two astronauts (who subsequently flew the first and second actual lunar landings) solely to determine the best hand-controller authority for LM attitude-control system.

Development of the activation procedures of the passive LM in flight (the first such requirement in manned space flight) necessitated a comprehensive and detailed systems simulation provided only by the LMS. The LM integrated activation checklists were worked out and verified for the two navigation systems, two radar systems, and three propulsion systems.

The maintenance of a high degree of mission-simulator fidelity was emphasized. No part of the system was more critical in this respect than the simulated guidance and navigation system. The use of an interpretive approach to simulate the Apollo guidance computer on both the CMS and the LMS provided not only exceptional fidelity but also the ability to accept late software changes and mission-peculiar tapes and to introduce these rapidly into the training. Numerous full-dress rehearsals were executed with the flight crews in the mission simulators at Cape Kennedy and the flight control team in the Mission Control Center at MSC. This training used the latest possible operational trajectory and spacecraft data to ensure complete verification of both flight and ground programs and comprehensive familiarization by the entire flight operations team.

Many simulations were executed with trainers other than the spacecraft mission and part-task simulators. Zero-g simulation of EVA contingency transfer by using the water-immersion facility at MSC (fig. 4-1) was an extension of the experience and techniques derived during Gemini. The partial-gravity simulator, which used a servoed, vertically mounted suspension system to take up five-sixths of the weight of the suited crewman, permitted long-duration 1/6g simulations. A truck-mounted device which used this same simulation technique allowed training at various traverse rates (fig. 4-4).

Many training hours went to one-g walk-throughs with mockups of spacecraft and equipment. Specific areas of training ranged from onboard stowage, photographic equipment handling, and docking-tunnel probe and drogue operations to full-blown lunar-surface EVA walk-throughs. The lunar-surface walk-throughs included special 1/6g mockups of the Apollo lunar-surface experiment package and the lunar-surface tools.

FLIGHT PLANNING

Apollo flight planning involves a much more complex procedures-integration task than did Mercury or Gemini, because of the large number of crew tasks associated with two manned spacecraft. Four or five months before launch, the flight planner must take the list of mission objectives, the preliminary launch-vehicle trajectory, the preliminary operational trajectory, the crew work and sleep constraints, and network tracking constraints and must integrate the numerous systems procedures of the many-phased Apollo mission. Because of the requirement to minimize use of reaction control propellant, hydrogen, and oxygen, several iterations are required to establish the proper timing of maneuver rates and spacecraft power configurations during the mission.

Original time line estimates for the various mission phases derive from the mission simulators in which all spacecraft systems are dynamically simulated. Spacecraft housekeeping, EVA preparation, and post-EVA procedures are worked out in high-fidelity mockups. The spacecraft stowage list itemizes 445 pieces of equipment, each of which must be procedurally integrated into the flight plan. Some 1300 copies of the flight plan are distributed to NASA and contractor personnel 3 months before the launch. If the flight plan is considered in a broad context to include all onboard paper including time lines, checklists, and graphics, the package weighs about 20 pounds. For each Apollo mission, hundreds of changes were made and verified on these data during the last 3 months before flight. Many changes reflected differences in the final trajectory, procedural refinements derived from simulation experience, and correction of anomalies on the previous flight.

We set a rigorous procedures change discipline for 2 months before the flight. This controls suggested alternate techniques which may appear preferable for a particular subsystem, but must be considered in terms of the interrelated operational time line before acceptance. We also establish a rapid procedures information loop during the final simulation phase, which also begins 2 months before the launch. This simulation activity involves flight planners, flight crews, mission simulators, flight controllers, and the tracking network. Normal and emergency procedures are thoroughly tested.

The Apollo flights speak for the value of this simulation effort in verifying late changes, validating procedures, and establishing crew readiness.

5. FLIGHT CONTROL IN THE APOLLO PROGRAM

By Eugene F. Kranz and James Otis Covington
Manned Spacecraft Center

Apollo 12 lifted off the pad at 11:22 a.m. e.s.t. on November 14, 1969. At 36.5 seconds after lift-off, lightning struck the command and service module (CSM), disconnecting all three fuel cells from the main buses and placing the main loads on two of the three batteries which ordinarily supply reentry power (fig. 5-1). Fuel cell disconnect flags popped up, and caution and warning lights winked on to alert the crew. With the decrease in the main bus power, the primary signal conditioning equipment ceased operating as it is meant to do when main bus voltages drop to approximately 22 volts. The ground simultaneously lost telemetry lock. At first, flight controllers thought the plume of ionized rocket-exhaust particles had blacked out the telemetry signal. However, they abandoned this theory when the crew reported the warning lights.

The primary signal conditioning equipment controls most electrical-power measurements; therefore, there was little information with which to diagnose the trouble. At 52 seconds after lift-off, the crew reported losing the spacecraft platform. At 60 seconds, the ground locked on to the telemetry signal again, and the CSM electrical and environmental systems engineer, John W. Aaron, asked the crew to switch to the secondary signal conditioning equipment to get additional insight into the electrical system. At 98 seconds, the crew made the switch, restoring all telemetry. Aaron then noted from his data display that three fuel cells were disconnected and requested the crew to reset them. Fuel cells 1 and 2 went back on the line at 144 seconds; fuel cell 3, at 171 seconds. Main bus voltages rose to approximately 30 volts, and all electrical parameters returned to normal.

Throughout the entire launch the Saturn launch vehicle performed normally. The spacecraft entered the proper orbit, and the crew and ground began preparing for translunar injection.

The quick response to the Apollo 12 outage came about not as a result of blind luck but of careful planning, training, and development of people, procedures, and data display techniques by those responsible for flight control.

The flight control organization devotes a majority of its time and resources to careful premission planning and detailed training. This premission preparation culminates in simulations of critical phases of the mission with the flight crew. These simulations prepare the flight controllers and the flight crew to respond properly to both normal and contingency situations.

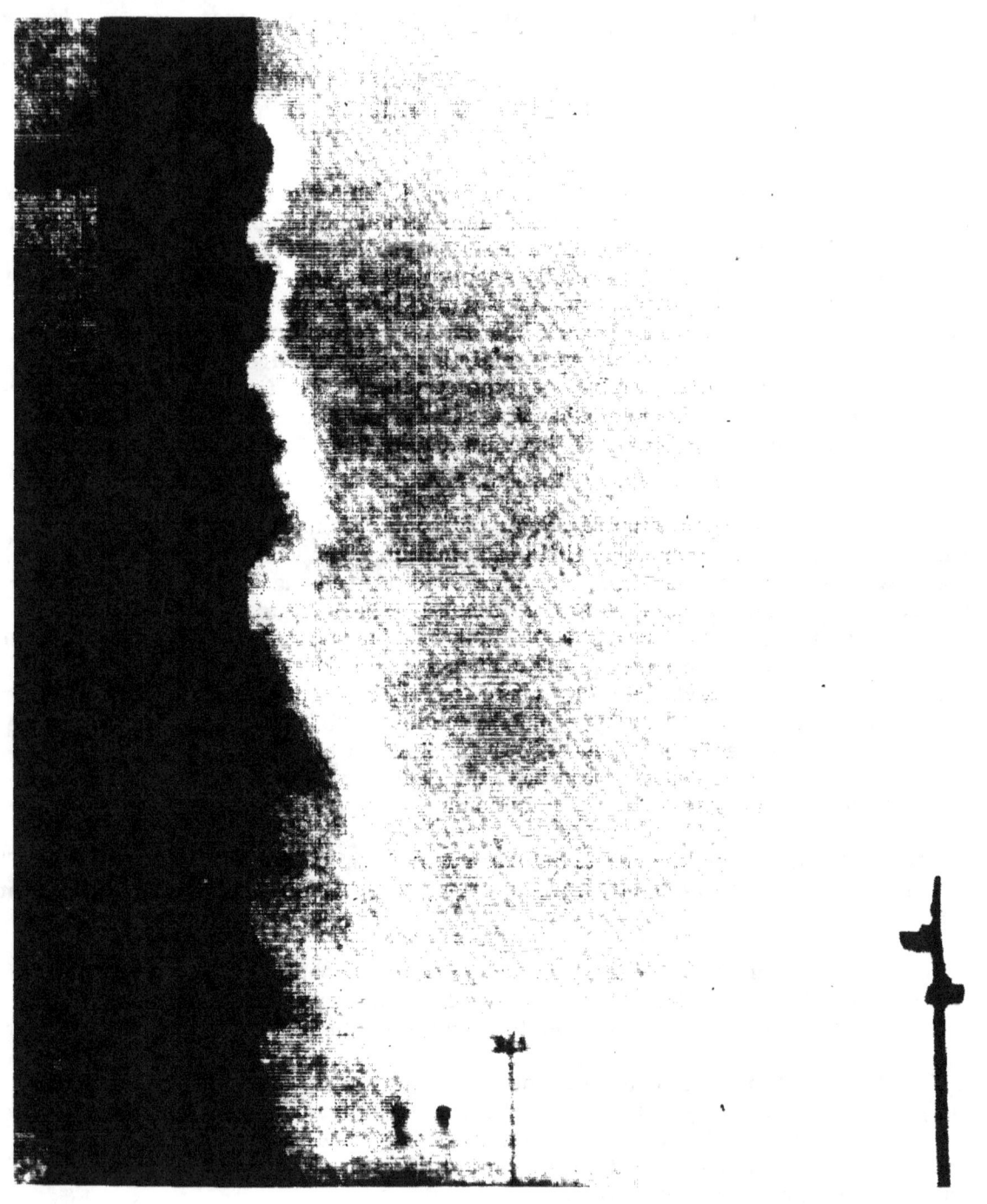

Figure 5-1.- Electrical power display when Apollo 12 was at an altitude of 6000 feet.

Following one of the basic Flight Control Division philosophies, operations personnel take part in planning a mission from its conception through its execution. They participate in two areas where their operational experience contributes greatly: (1) in the early stages of mission design and (2) in setting basic design requirements for various spacecraft systems. As a result, both spacecraft hardware and mission design have optimum operational qualities.

The true operational phases of the Flight Control Division begin after the spacecraft design reaches completion and NASA has committed itself to constructing flight vehicles and launch facilities. The development of mission operations starts after sufficient information about the detailed design of the space vehicles becomes available. The development takes 2 years and is divided into four phases, as shown in figure 5-2: mission development, detailed planning, testing and training, and real-time operational support of a flight.

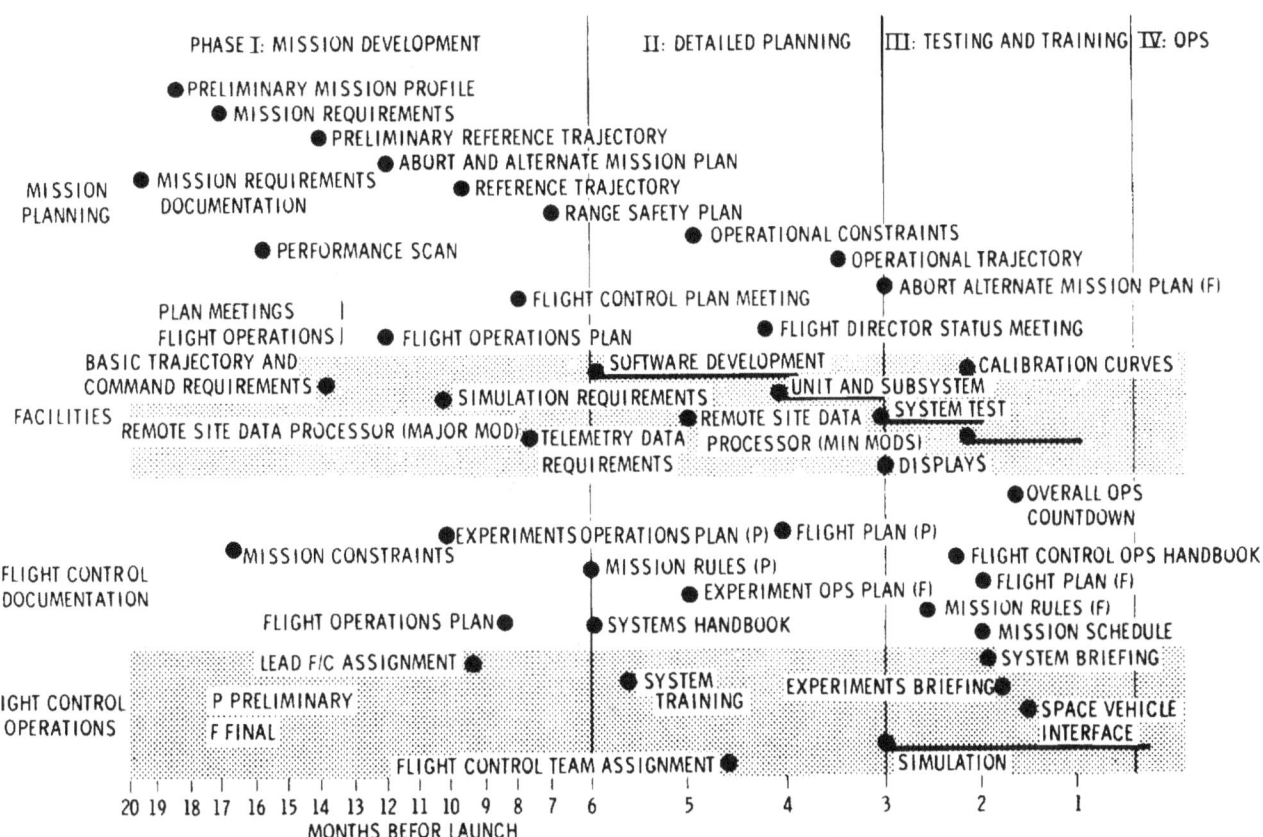

Figure 5-2. - Mission-development time line.

The mission development phase begins approximately 2 years before the first launch. First comes the establishing of the conceptual operational guidelines for the flight. The guidelines for the Apollo Program were developed during Project Mercury and the Gemini Program.

Throughout the Mercury and Gemini flight programs, teams of flight controllers at the remote tracking stations handled certain operational responsibilities delegated to them somewhat independently of the main control center. As the flight programs progressed, the advantages of having one centralized flight control team became more apparent. By the advent of the Apollo Program, two high-speed (2.4 kbps) data lines connected each remote site to the Mission Control Center (MCC), permitting the centralization of flight control there.

The spacecraft systems were of two functional categories: (1) electrical, environmental, and communications and (2) guidance and control and propulsion. A flight controller in the Mission Operations Control Room (fig. 5-3) of the MCC monitored each one. He received the backing of the Staff Support Room where every spacecraft system had its own man.

The flight dynamics team philosophy remained unchanged from earlier programs. The team would monitor the spacecraft trajectory and plan changes to it, monitor and manage the three spacecraft computers, and plan the return to earth.

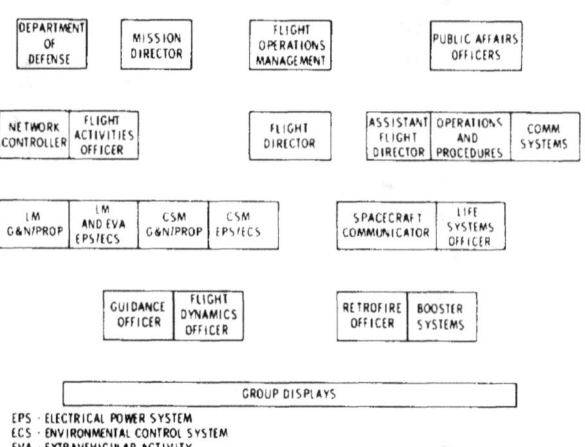

Figure 5-3.- Mission Operations Control Room divisions.

The function of the Flight Director and his staff also remained unchanged. The Flight Director directs and coordinates the flight control team. He may, after analysis of the flight, take any action necessary to complete the mission successfully.

Following mission definition, flight control personnel participate in flight operations planning meetings. At this intercenter (Manned Spacecraft Center, Marshall Space Flight Center, Kennedy Space Center, Goddard Space Flight Center, and NASA Headquarters) forum, operations personnel can directly influence the first detailed mission design. As the mission design begins to take shape, they have an opportunity to analyze the possible effects of various failures on the successful completion of the mission. They can then suggest alternate mission designs or hardware changes (or both) which would improve chances for success. As the mission plan solidifies, the operations personnel identify certain guidelines and constraints forming boundaries within which the detailed mission planning will take place.

Concurrent with the development of the operations concepts and the mission guidelines and constraints, systems flight controllers begin to gather detailed systems information from the spacecraft manufacturers. The systems flight controllers get functional schematics and engineering drawings which they translate into a handbook of spacecraft systems. Flight control personnel and the flight crew use this handbook during the mission. On one page, like that reproduced in figure 5-4, each schematic shows all system or subsystem interfaces, together with the power sources, onboard and ground instrumentation, the various controls and displays necessary for the operation of that system, and pertinent notes on system performance. As a spinoff, preparation of the systems handbooks provides a chance to spot potential systems design

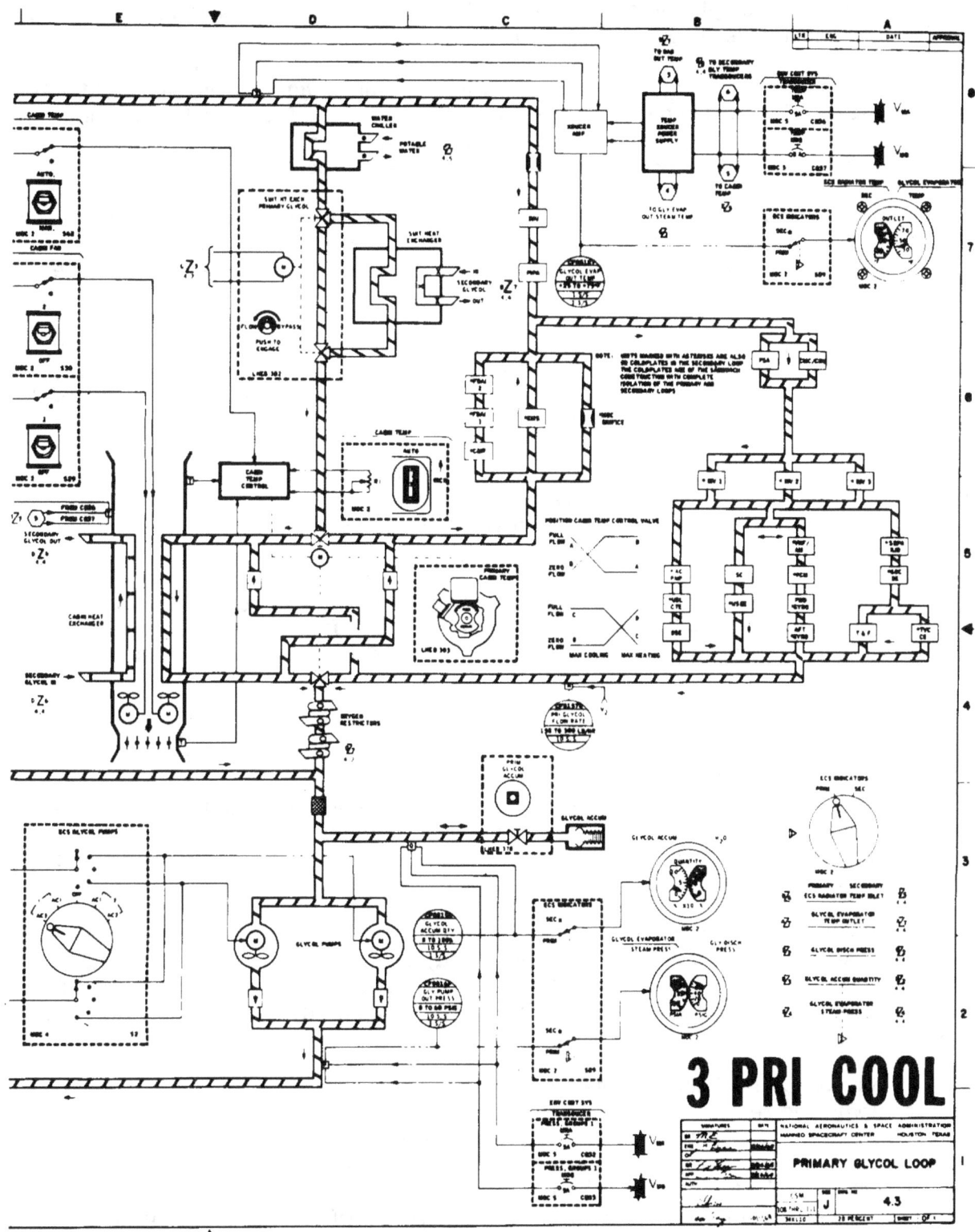

Figure 5-4. - Partial sample of CSM systems schematic.

inadequacies. On completion of the systems handbooks, the long-lead-time items comprising the mission development phase come to a close.

The phase II detailed planning begins 6 to 9 months before launch when NASA assigns detailed mission objectives to the particular mission. The objectives are assigned to a particular phase in the mission time line, and flight control personnel work with the Apollo Spacecraft Program Office to establish their order of priority.

Flight controllers begin defining the specific data necessary to carry out the detailed mission objective and monitor the operational spacecraft systems. They do this parameter by parameter, assigning a priority to each. Programers can then write the software required to provide these data to the flight controllers in a usable form.

Meanwhile, the mission rules begin to take shape. The rules specify in great detail how to conduct the mission in both normal and contingency situations. The final list constitutes a three-way agreement among flight control personnel, the flight crew, and management personnel.

Flight crew safety overrides all else. Then come into play complex tradeoffs between mission objectives and spacecraft design, the reliability and maturity of all elements associated with the conduct and control of the mission, the mission-objective priorities, and risk tradeoffs.

Mission rules fall into two distinct categories: (1) general guidelines formulated by the Flight Director and his staff and (2) detailed rules formulated by individual flight controllers in response to the general guidelines. One general guideline reads as follows:

During lunar module powered descent, if a systems failure occurs, a choice is available.

1. Early in powered descent when descent-propulsion-system-to-orbit capability is available (up to powered descent initiation plus 5 minutes), it is preferable to abort in flight rather than to continue descent. Redundant capability of critical lunar module systems is required to continue powered descent during this period.

2. During the remainder of powered descent, it is preferable to land and launch from the lunar surface rather than to abort. Only those system failures or trends that indicate impending loss of the capability to land, ascend, and achieve a safe orbit from the lunar surface, or impending loss of life-support capability will be cause for abort during this period.

This rule came about because up to 5 minutes after powered descent initiation, one can abort and reach orbit by using only the descent stage. One can keep the descent stage and use its consumables (oxygen, water, and batteries) if it takes a long time to rendezvous or if one loses the ascent consumables. After 5 minutes in powered descent, the descent propellant remaining could not return the craft to orbit. It then becomes more desirable to continue on and land. By landing and lifting off one revolution (2 hours) later, one gains sufficient time to analyze the malfunction, perform any system reconfigurations necessary, and perform a nominal ascent and rendezvous. Since the flight crew and the flight controllers emphasize nominal activities most, they know them best and always use them if possible.

Detailed rules, such as the ones listed in figure 5-5, expand upon the broad philosophy of the general guidelines to cover single failures of all individual systems and subsystems throughout the spacecraft down to the individual instrumentation points necessary to making mission-rule decisions.

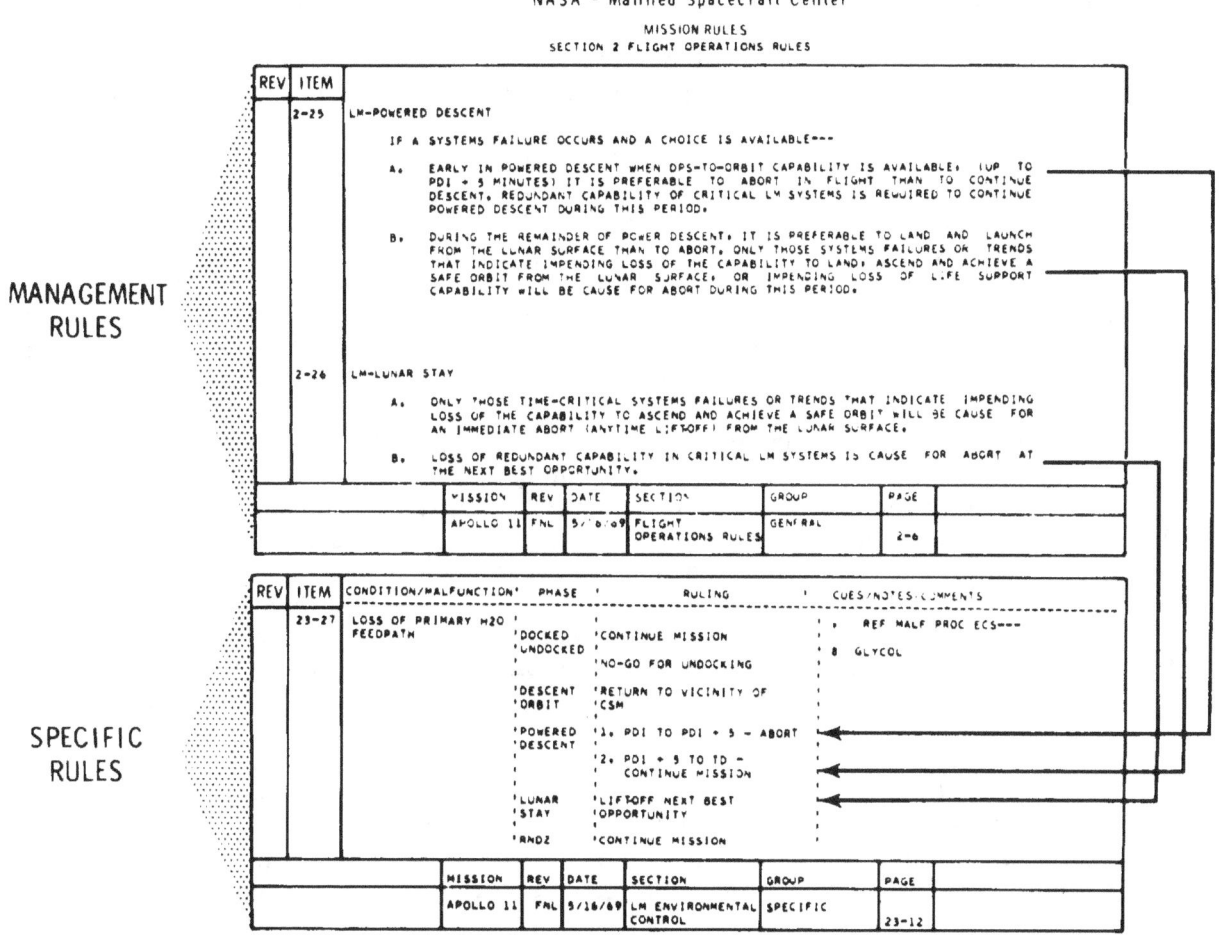

Figure 5-5. - Sample of flight mission rules.

The specific format for mission rules has remained the same since Project Mercury. A short statement describes the condition or malfunction which may require action. The ruling follows. A third section of the rule contains applicable notes, comments, or standard operating procedures. Documents carefully delineate the reasoning behind each mission rule, including tradeoffs which may not be otherwise apparent.

Standard operating procedures receive as careful attention. They divide into interface procedures and personal procedures. Interface procedures include all flight control procedures involving more than two console positions. Together, the procedures comprise the Flight Control Operations Handbook (FCOH). Like the mission rules, the FCOH procedures appear in a standard format and are as brief as possible without becoming ambiguous. Checklists, like the one shown in figure 5-6, are used whenever possible.

Personal procedures go into the Flight Controller Console Handbook prepared for each console position. This handbook describes the job of the flight controller manning that position and details the procedures he must follow. As does the FCOH, it uses, whenever possible, checklists of the kind shown in figure 5-7.

The testing and training phase has two main purposes: (1) integrating the flight control team and (2) testing the procedures and mission rules prepared for the specific mission. Six distinct techniques are used to train flight controllers for a mission. Several are used simultaneously. As described earlier, documentation development proved to be one of the most successful techniques. The system flight controllers gain an intimate knowledge of the hardware design and what it will and will not do through their long weeks and months of preparing the spacecraft systems handbooks. Likewise, the flight controllers learn the various ramifications of the mission rules by discussing the various merits of alternate courses of action. Preparing the procedural documents instills in the flight controller the capabilities and responsibilities of his position and the methods of using his position most efficiently.

Formal classes convene periodically to examine spacecraft systems in detail, to present basic orbital mechanics laws and principles, and to give instruction on the capabilities and limitations of the ground systems. Programed instruction methods, as developed in 1965 by the Air Force, underwent slight modification for flight control training. Programed-instruction courses covered all spacecraft systems and, ultimately, all functional job positions in various levels of detail, thus providing two benefits. The courses provided easily assimilated material, such as that shown in figure 5-8, which helped individual flight controllers understand the problems and responsibilities of other flight controllers and assisted in training controllers who could move rapidly from one job to another.

During the Gemini and Apollo Programs, cockpit-system trainers were developed for flight control training. These automated trainers allow the flight controllers to become familiar with the problems encountered by the flight crew. With the trainers, flight controllers identify, develop, and exercise critical crew procedures.

Through his support of program office design reviews, single-point failure reviews, and detailed mission requirement reviews (to name a few), the flight controller could associate directly with the designers, builders, and operators of the flight hardware. This often provides a rare insight to critical information that may affect all areas of planning and preparation of a mission.

Simulation began 2 or 3 months before launch. The entire flight control team took part. Artificially created spacecraft and network data flowed into the control center, providing real-time responses and displays as if a mission were actually in progress.

The first mission simulations worked with computer-generated mathematical models of space vehicles. These provided flight controller training in exercising documented procedures and mission rules in response to live data and a dynamic mission situation. Approximately 30 percent of the total simulation time goes into working with the mathematical model.

SOP 4.1

REV A

TM FORMAT SELECTION

I. PURPOSE

THE CAPABILITY EXISTS TO SELECT 19 TM FORMATS FROM A HSD SITE. THE FOLLOWING PROCEDURES DEFINE UTILIZATION OF THE FORMATS AND ASSIGN PRIORITIES TO VARIOUS FORMATS DEPENDING ON MISSION PHASE.

II. PARTICIPATION

TIC	BOOSTER
GUIDANCE	HOUSTON TM
PROCEDURES	G&N
TELMU	CONTROL

III. PROCEDURES

A. **MSFN HSD FORMATS**

1. TABLE 4.1-2 SHOWS THE NOMINAL HSD FORMAT PLAN. THIS PLAN PROVIDES TIC WITH THE HSD FORMAT NUMBERS TO BE TRANSMITTED TO THE SITES BY F-2 DAYS. THE STATION M&O WILL BRING HIS SITE UP IN THE REQUIRED FORMATS BY H-15 MINUTES. FOR DEVIATIONS FROM THIS PLAN, PROCEDURES WILL PROVIDE TIC THE TM FORMAT NUMBER TO BE EXECUTED VIA PRM AT H-5 MINUTES TO THE SITE'S PASS.

2. IF AN MCC F/C DESIRES TO CHANGE THE HS TM FORMAT, HE WILL COORDINATE THROUGH PROCEDURES. PROCEDURES WILL COORDINATE WITH OTHER AFFECTED F/C'S AND IF NO OBJECTIONS ARE RECEIVED ON AFD CONF LOOP WITHIN 15 SECONDS, PROCEDURES WILL REQUEST THE CHANGE FROM TIC ON THE CCATS TM MON LOOP. TIC WILL ISSUE A PRM FORMAT CHANGE EXECUTE FROM HIS CONSOLE. IF TIC HAS NO INDICATION OF FORMAT CHANGE WITHIN 5 TO 10 SECONDS, TIC WILL REISSUE THE PRM EXECUTE. IF TIC HAS NO INDICATION OF FORMAT CHANGE AFTER THREE EXECUTES, HE WILL REQUEST THE STATION M&O ON GOSS 4/NET 2 TO MANUALLY SELECT THE DESIRED FORMAT.

3. DURING THE LAUNCH PHASE, BOOSTER AND/OR GUIDANCE CAN DIRECT TIC TO CHANGE HSD FORMAT CHANGES ON THE CCATS TM MON LOOP.

B. **MCC PCMGS ANALOG PATCHING FORMATS**

G&N (VEH SYS SSR), AFTER PROPER COORDINATION IN THE SSR, WILL DIRECT HOUSTON TM DURING THE PROPER MISSION PHASES OVER THE RTCC TLM LOOP TO SWITCH THE FOLLOWING PCMGS FORMATS.

FORMAT	WHEN USED
9	LAUNCH TO LM POWERED PHASE
8	LM COAST PRIOR TO DPS MANEUVER, DOCKED LM/CSM, CREW TRANSFER, EMU ACTIVITY
7	LM DPS MANEUVER, CSM COAST
6	LM APS MANEUVER, CSM COAST
5	LM COAST AFTER APS MANEUVER, CSM SPS MANEUVER
4	CSM ANALYSIS

Figure 5-6. - Sample of Flight Control Operations Handbook.

SCP NO. **ECS 6-3**

S/C: LM
DATE: 3/27/69
REV: Original
ORIGINATOR: Smith
APPROVAL: *[signature]*

TITLE: O_2 QUANTITY CALCULATIONS

PURPOSE: This SCP describes the method for determining O_2 quantities in pounds from telemetered pressure and temperature.

PROCEDURE:
A. General - The mass of a body of gas can be determined by using the modified gas law formula

$$M = \frac{KP}{T_R Z}$$

where M = mass to be determined in pounds

K = 8.91 for descent tank
  = 1.48 for a single ascent tank

P = Pressure in psia (GF3584P, GF3583P, GF3582P)

T_R = (TC + 460)

Z = Compressibility Factor obtained from Table 1 or from a chart of generalized compressibility factors. Z is inserted into the RTCC as a MED.

B. The LM ECS Engineer will insert the compressibility factor from TABLE 1 into the RTCC as required. The factor for 70°F is considered adequate for the expected range of temperatures.

TABLE 1 - O_2 QUANTITY CALCULATION INPUTS

DESCENT TANK

PRESSURE (PSIA)	QUANTITY (POUNDS) at 60°F	at 70°F	at 80°F	COMPRESSIBILITY (Z) at 70°F
2900	53.92	52.44	51.47	0.930
2700		48.83		0.930
2500	46.48	45.16	43.85	0.931
2300		41.42		0.934
2100	38.75	37.73	36.64	0.936
1900		33.99		0.940
1700	31.00	30.22	29.38	0.946
1500		26.44		0.954
1300	23.39	22.77	22.24	0.960
1100		19.15		0.966
900	15.94	15.57	15.21	0.972
700		12.00		0.981
500	8.66	8.49	8.33	0.990
300		5.05		0.998
100	1.71	1.68	1.65	1.00

ASCENT TANK

PRESSURE (PSIA)	QUANTITY (POUNDS) at 60°F	at 70°F	at 80°F	COMPRESSIBILITY (Z) at 70°F
900	2.64	2.58	2.52	0.972
700		1.99		0.981
500	1.44	1.41	1.38	0.990
300		0.84		0.990
100	0.28	0.28	0.27	1.00

Figure 5-7. - Sample of Flight Controller Console Handbook.

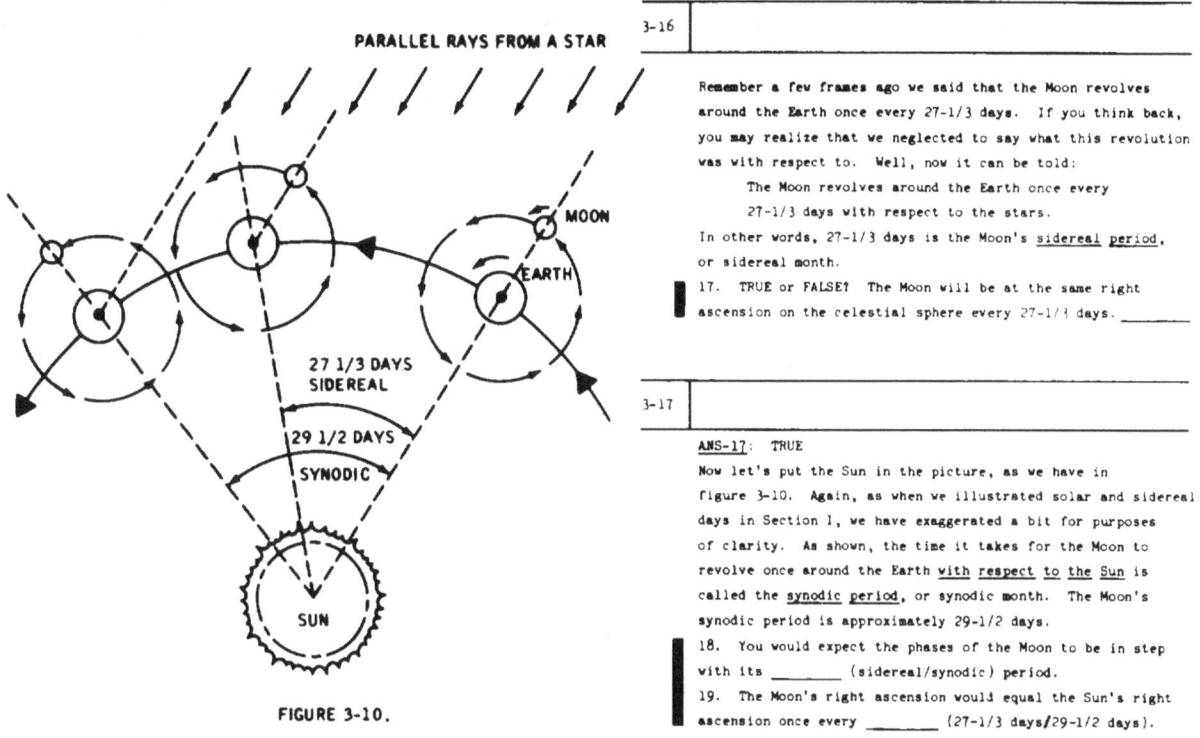

Figure 5-8. - Sample of programed-instruction text.

During the remainder of the time, controllers worked with the flight crew spacecraft mission simulators. This promotes the integration of the flight control team and the flight crew team. Simulation personnel introduce spacecraft and Manned Space Flight Network problems and failures to test the documented procedures and mission rules. Sometimes the simulations lead to changes in them. The time spent on each period depends on its criticality and the amount of past experience available.

For an Apollo mission, simulations cover these periods of major activity: launch through translunar injection; lunar orbit insertion; lunar module activation, checkout, and descent; lunar surface operations; ascent through rendezvous; transearth injection; and reentry.

We found in the Apollo Program that having groups specialize in these periods gave better support. Thus, four teams divide the periods among themselves. Each gets to know critical mission phases better, but does not have to spend such long hours on the simulators.

The culmination of all this preparation follows lift-off. Throughout the Apollo Program, approximately 80 percent of all problems encountered in flight, whether large or small, had been previously discussed, documented, and simulated before the flight. This made choosing the correct course of action almost automatic. The remaining 20 percent of the problems readily yielded to the same logic, and decision-making procedures followed to arrive at premission decisions. The logic of flight control decisions is diagramed in figure 5-9.

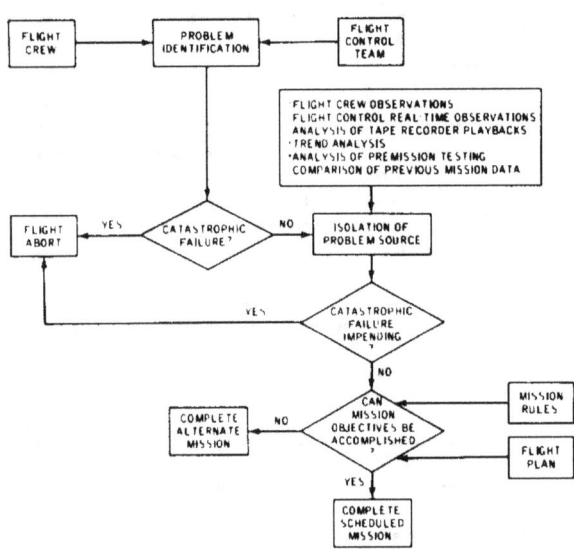

Figure 5-9. - Logic of flight control decisions.

A functional organization has emerged which is flexible enough to meet unexpected problems, but is structured enough to provide continuity of operation from mission to mission. The basic principles of flight control are not unique to manned space flight. They apply to any field where one can visualize malfunctions, document solutions, and rehearse the resulting actions. They could find use in any field where one monitors equipment or procedures by remote sensing devices. Application of the basic principles could increase efficiency in any field where one can write standard operating procedures. In the Apollo Program, they helped carry man to the lunar surface and bring him safely home again.

6. ACTION ON MISSION EVALUATION AND FLIGHT ANOMALIES

By Donald D. Arabian
Manned Spacecraft Center

The aim of mission evaluation is to extract the maximum amount of information from each flight for use in managing and planning future missions and programs. Its products also go before the scientific community and the public.

It is not surprising that a complex vehicle, such as a spacecraft, experiences peculiarities and system problems that have not been considered previously. The prime task of the mission evaluation team during a mission, therefore, involves identifying and understanding these peculiarities and problems and determining what action should be taken. After a mission, the evaluation team also has responsibility for (1) flight anomaly investigation and resolution and (2) preparation and publication of the mission report.

For the mission evaluation, the many engineering disciplines involved in Apollo are represented by groups of specialists, each managed by a NASA team leader. In general, the specialists have experience with a particular system from the initial design conception through development and testing of the hardware. Consequently, they have an intimate knowledge of the system operation and limitations. The specialists are selected both from contractor and from NASA engineering organizations and work as an integrated team. Before a mission, the selected team participates in certain simulations with Mission Control Center personnel. This training serves to integrate the two organizations into a single operational unit.

The responsibility of each team leader includes coordination with other team leaders to ensure that any solution or recommended course of action for his system does not jeopardize other systems. The team leaders report to the evaluation team manager, who is responsible for the overall operation of the team. The team manager is assisted by senior contractor engineering managers who have immediate access to the company facilities to provide any necessary support. All inputs and requests by the team leaders are integrated and reviewed by the team manager and by contractor engineering management. The team manager interfaces with the Flight Director through the Apollo Spacecraft Program Manager. During a mission, the team mans assigned positions in the mission evaluation room (fig. 6-1) around the clock.

To support the mission evaluation team effort at the NASA Manned Spacecraft Center (MSC), the prime contractors maintain similar teams of specialists at their facilities for analyses, tests, and related activities. These contractor teams are also under the direct management of the mission evaluation team at MSC.

A typical example of the activity associated with these supporting teams occurred when a lightning discharge on Apollo 12 caused the loss of inertial reference (tumbling) of the inertial platform. In this instance, the support team at the Massachusetts

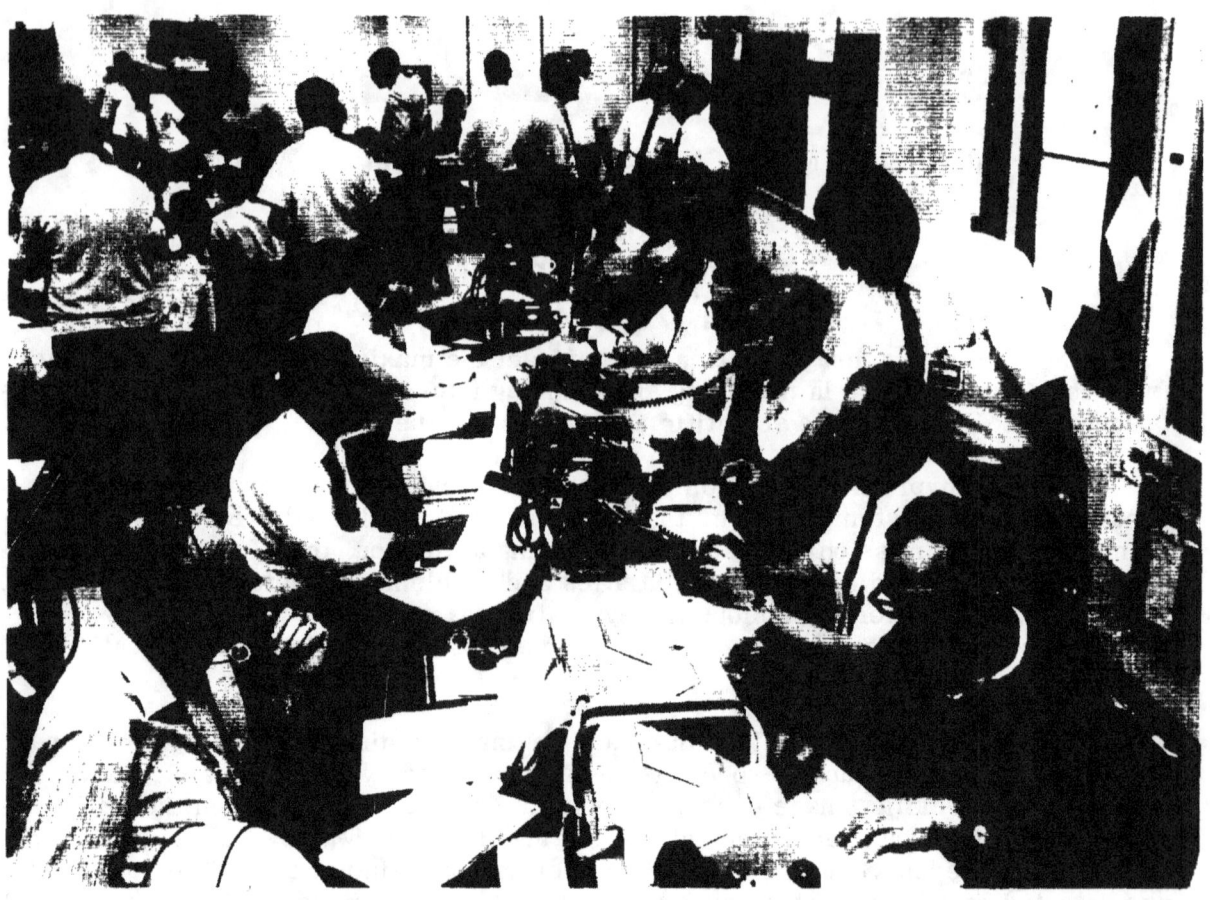

Figure 6-1.- Mission evaluation room with team leaders' table in the foreground and discussion of a system problem in the background.

Institute of Technology performed a test to simulate the conditions so that it could be understood how the potential discharge caused the platform to tumble. It was important to have rapid verification that the platform had not been damaged. A quick response was received by the evaluation team, and the mechanism that had caused the conditions in flight was determined.

Many other times during Apollo flights, the evaluation team has provided test information that was valuable in understanding a problem and determining the best course of corrective action. In carrying out its task, the team must be constantly aware of the total system performance as the mission progresses. The team monitors data that are received by the tracking stations, transmitted to and processed by the Mission Control Center, and displayed in the Mission Evaluation Room on closed-circuit television. In addition, air-to-ground voice communications are monitored.

The frequency of the Apollo flights demands that anomalies be quickly identified and resolved so that prompt corrective action may be taken. Analysis of the data for problems and anomalies must be compressed, therefore, into a relatively short period. Also, within this time frame, the anomalies must be analyzed to the extent that the mechanism associated with the cause is clearly understood. Of course, anomalies which involve flight safety or which would compromise the mission require corrective action before the next flight.

The first problem is to identify the anomalies. Many anomalies are easily recognized because a component has failed to operate. The most difficult cases, however, appear when the data from the system are not sufficient for an understanding of all the normal operating characteristics. A typical example of this condition occurred during the Apollo 7 mission when the battery recharging characteristics fell below predicted levels throughout the flight. Preflight tests had been conducted at the component level, but an integrated test of the entire system, as installed in the spacecraft, had not been included. Postflight tests, using the actual flight hardware, showed the same characteristics as those experienced in flight. A detailed analysis showed that the line resistance between components of the system greatly controlled the amount of energy returned to the battery. The corrective action for this anomaly was to require that integrated system tests be performed to establish overall system characteristics of each installation and thus ensure adequate battery recharging. In this case, if the total system operating characteristics had been established previously, there would have been no problem.

Sometimes, also, data will not support an accurate analysis of a problem. This situation occurs because of insufficient flight instrumentation or absence of recorded data. The mission evaluation team, in attempting to focus on the anomalous condition, must rely on the history compiled from previous missions and on the experience gained from tests and checkout and knowledge of the failure history.

After an anomaly is identified, the cause and the corrective action must be identified. The approach may be experimental or analytical or both.

The depth and the extent of the analysis vary considerably and depend on the significance of the problem. For example, on the Apollo 6 mission, a structural failure occurred in the adapter that holds the service module to the launch vehicle and also houses the lunar module during the ascent portion of the flight. This adapter consists of the largest honeycomb structure designed and developed for any application. Long-range photographs show that the structure lost the outer face sheet from the honeycomb-sandwich panel by explosive separation. (See fig. 6-2.) Response to the effects was obtained from many measurements in the command and service module, lunar module, and launch vehicle.

In resolving this problem, four possible causes were investigated: structural dynamics of the launch vehicle, the dynamic loads of the lunar module, dynamic modes of the adapter shell itself, and quality of manufacture.

The investigation first focused on an understanding of the coupled vibration modes and characteristics of the launch vehicle and spacecraft. Extensive vibration tests ruled out vibration as a cause.

Further tests and analyses indicated that the internal pressure of the sandwich panels could have caused the failure. If a large unbonded area existed between the honeycomb and the face sheets, then aerodynamic heating of the air and moisture entrapped

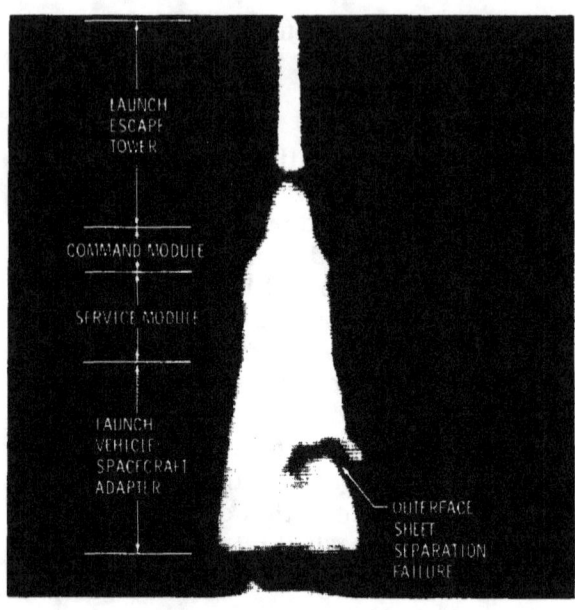

Figure 6-2.- Long-range photography of adapter failure during the Apollo 6 mission.

in the panel could have caused a pressure buildup in the honeycomb, separating the face sheets in an explosive manner. The most probable cause of this condition was traced to the manufacturing process (the possibility of unbonded face sheets remaining undetected). To circumvent the problem in the future, the inspection procedures for the structure were changed. Also, cork was placed on the outer surface of the adapter to reduce aerodynamic heating, and small vent holes were drilled through the inner face sheet into the honeycomb to reduce internal pressure.

Analysis of this anomaly involved testing full-size equipment under dynamic and static conditions, performing many experimental tests of smaller test articles, and conducting extensive structural analysis at the various NASA centers and at many contractor organizations. This effort verified that the structural integrity of the adapter was satisfactory for subsequent missions and established that the failure was not caused by a design deficiency.

The other extreme of treating a problem concerns anomalies for which no corrective action is taken because of the nature of the failure. For example, on the Apollo 11 entry-monitor system, an electroluminescent segment of the velocity counter would not illuminate. A generic or design problem was highly unlikely because of the number of satisfactory activations experienced up to that time. A circuit analysis produced a number of mechanisms that could cause this failure, but there was no failure history in any of these areas. This case is a typical example of a random failure. The basic design concept of the spacecraft overrides such failures by providing alternate procedures or redundant equipment. Consequently, this type of failure does not demand corrective action.

Causes of anomalies involve quality, design, and procedures. The quality items include broken wires, improper solder joints, incorrect tolerances, improper manufacturing procedures, and so forth. The structural failure of the adapter on the Apollo 6 mission, previously discussed, illustrates such a quality problem.

System anomalies caused by design deficiencies can generally be traced to insufficient design criteria. Consequently, the deficiency can pass development and qualification testing without being detected, but will appear during flight under the actual operational environment. A typical example of a design deficiency is the fogging of the Apollo 7 command module windows between the inner surfaces of the three window panes. A postflight examination showed the fogging to be a product of the outgassing of the room-temperature-cured sealing material used around the window. The design criteria did not require the sealing material to be cured; curing would have prevented outgassing in the operating temperature and pressure environment.

Procedural problems in operating various systems and equipment are ordinarily corrected simply. For example, an incorrect procedure was used to chlorinate the crew's drinking water. This resulted in an improper mixing of the chlorine and water, giving the water a strong chlorine taste. The procedure was revised, and the water no longer had an objectionable taste.

While a mission is in progress, the scope of a problem must be defined before action is taken; that is, the problem must be isolated to the extent of establishing possible effects on the spacecraft systems. Every effort is concentrated on developing a procedure that permits the mission to continue despite the problem. Complete understanding and analysis of the problem frequently require postflight data analysis, crew debriefings, or testing of the spacecraft hardware.

When lunar module hardware exhibits a problem, it may be returned for postflight analysis and testing, provided the command module has stowage space available and the lunar module gear fits the space. For example, the color television camera that failed on the lunar surface during the Apollo 12 mission was scheduled to be left on the lunar surface but was returned to earth for postflight testing.

When postflight testing cannot be done because the hardware cannot be returned to earth (e.g., the lunar module and service module hardware), every effort is made to perform inflight tests to understand and isolate the cause of the failure. A typical case occurred during the Apollo 12 mission when the high-gain antenna, mounted on the service module, occasionally exhibited reduction in signal strength. The evaluation team developed a test to isolate the anomaly to specific areas and components of the high-gain-antenna system. The test, conducted later in the mission, produced data which, combined with analysis, isolated the cause.

An additional search for anomalies is conducted when the command module is returned to the contractor's facility for a general inspection. Those systems and components identified as having a problem or failure are removed from the vehicle and tested to establish the cause, or tests are performed with the affected equipment in position in the command module. In general, these postflight tests are limited to those required for an understanding of flight problems.

The concerted effort on anomalies during the flight continues after the mission until each problem is resolved and corrective action is taken. This activity requires close coordination and cooperation between the various Government and contractor groups. Emphasis is placed on prompt and exact analysis for the understanding and timely solution of each problem.

To accomplish this task, a problem list is maintained during and after each flight. This list contains a discussion of each problem, the action being taken to resolve it, the engineer or contractor responsible for completing the action, and the anticipated closure date.

After the flight, the most significant anomalies are published in the 30-Day Anomaly and Failure Listing Report. This report identifies the anomaly, discusses the analysis, and identifies the corrective action to be taken. The Mission Report, which is published approximately 60 days after a flight, includes a section which discusses all significant anomalies and corrective actions. The less significant problems are discussed in the appropriate sections of the Mission Report. A separate report covers anomalies not resolved before publication of the Mission Report. The Mission Report serves as an historical record of the pertinent events of the mission and includes discussions of systems performance, crew activities, flight anomalies, and scientific experiments. Supplements to the Mission Report are published to present detailed analysis of the engineering systems, medical aspects, and scientific aspects of the mission.

The techniques used in Apollo for assessing systems performance reflect significant advancements over those used in previous manned programs. The method of handling flight anomalies, including the depth and extent of analysis, has been sufficient for the time and economy constraints imposed by the program. The Apollo concept has proved to be very effective in organizing many contractors and federal organizations into one central team for the real-time support and postflight evaluation of each mission. These concepts enabled the Apollo Program to advance at the rate required to achieve the national goal of landing man on the moon before 1970.

7. TECHNIQUES OF CONTROLLING THE TRAJECTORY

By Howard W. Tindall, Jr.
Manned Spacecraft Center

Someone not associated with the Apollo Program cannot imagine how much planning precedes each mission. Planning is truly an immense task, which takes many different forms. Basically, this is another report about mission planning. It is not about the selection of launch windows, trajectories to be flown, landing sites, or how to make the lighting conditions right. Neither does this report specifically involve preplanning the crew time line; defining when pieces of equipment should be turned on, when the crew should sleep, eat, work, perform on TV; and so forth. The planning this report is concerned with does interact intimately with these, but it is planning of an entirely different type. Generally speaking, it is the planning required to define how the trajectory is controlled once the mission objectives, trajectory plan, and crew time line have been established: to figure out exactly how the various components of the guidance, navigation, and control (GNC) systems and, to some extent, the engines are to be used during all phases of each of the manned Apollo missions.

Unquestionably, activities associated with trajectory control constitute by far the largest piece of operational overhead in any Apollo mission. That is, in the process of achieving the real objectives of a lunar-landing mission (such as placing experiments on the lunar surface, picking up rocks, and taking pictures) you will find no other inflight activity that approaches trajectory control in its capacity to absorb planning and training energy. Usually, when conflicts arise, the trajectory-control activity takes priority over everything else, such as systems management, crew work and rest cycles, and experiments (including the lunar-surface work).

Since trajectory-control activities make up such a large part of every Apollo mission, the manner of conducting them has an impact on almost every other facet of the mission, even on aspects that seem remote, such as electrical-power management, thermal control, and consumable budgeting. As a result, it is necessary to consider all of these things in the development of the overall trajectory-control procedures, which we call mission techniques. Also, it is important that almost everyone in the Apollo world knows how we intend to do these things, since we often impact their plans.

One basic characteristic in the design of the overall Apollo guidance and navigation (G&N) system must be recognized before I proceed. Namely, the ground-based tracking, computation, and control-center facilities not only form an integral part of the G&N system — being the prime source of essential data to the spacecraft systems — but in some instances, the ground will be the only source of data. Specifically, with the exception of rendezvous and inertial sensing during maneuvers, all trajectory determination is done on the ground. This is the task of determining the position and velocity of the spacecraft and its relation to where we are trying to go — braking into lunar orbit, trying to land at a specific site on the moon, hitting the earth atmospheric-entry corridor, or whatever.

Also, with the exception of the rendezvous, the ground is the only source of maneuver targeting. By maneuver targeting, I mean the task of figuring out the exact magnitude and direction and the time at which each maneuver must be executed in order to perform the mission.

The two essential functions, orbit determination and targeting, cannot be performed on board the spacecraft. The reason I am making such a big point about this is probably obvious. Putting together mission techniques with a G&N system like that is much more complicated that if the whole job could be done on board the spacecraft without external assistance. A tremendous amount of data must be relayed back and forth between the spacecraft and the ground, and the content and format of these data have to be complete and precisely compatible. Also, instead of only the three crewmembers being involved in the operation — that is, understanding and carrying it out — we must involve the entire flight-control complex. This makes the inflight job, of course, more complicated — but, believe me, it makes the planning job something else, too. Many diverse opinions about the planning task are expressed without hesitation or inhibition.

There is an important point I would like to make regarding operational philosophy, since it influences the mission techniques so much. It must be obvious from the way the missions have gone so far that the spacecraft equipment is really put together right. A tremendous amount of attention has been devoted to design and testing to make sure that everything will work as it is supposed to. On top of that, in order to provide even more confidence that the mission may be conducted successfully and safely, all of the critical systems are backed up by other systems. (In the GNC area, the backup systems, without exception, are of an entirely **different design than** the primary systems. Thus, they are operated entirely differently, which almost doubles the planning.)

Accordingly, you might think that the mission techniques for a spacecraft like Apollo would be based on the assumption that the equipment will work. The fact is, however — right or wrong — that our operational planning on all manned space flights so far has been based on a philosophy of cross-checking and monitoring every critical system and operation to make sure that the systems are performing properly.

Operations must all be planned out before a mission — just what will be done if the primary or backup system (or both) fails or is degraded. Also, we must plan what can be done with whatever components of these systems might still be working, with the data shipped from the ground and from charts and simple mathematical techniques developed for the crew, and with any other data sources not ordinarily used, such as the view out the windows.

On the other hand, when you consider that we deploy two sophisticated spacecraft, each of which has extensive capabilities, the imagination explodes with the possible ways things could be done as various components fail. Also, when you multiply this multitude of choices by the number of uniquely different phases in a lunar-landing mission, you find that the task of developing the techniques could expand indefinitely. Obviously, if unconstrained, this whole business could get out of hand. Not only could it cause an inexcusable waste of resources, but it could also introduce a myriad of complicated procedures — tier upon tier of alternate modes of operation which would go beyond the ability of the crew and flight controllers to understand, train with, and use.

These two rules bound the problem: Be prepared to recognize and react under any condition to save the crew and the mission, but do not carry this business to the point of actually reducing reliability by introducing confusion or the incomprehensible into the system. If I were to look back and judge how we actually did on Apollo, I would say we went a little too far — not much, but some. And, of course, it is easier to look back.

The Mission-Technique Development Task: How do you decide how to fly an Apollo mission? If the inertial reference system is drifting a little, what do you do about it? If it drifts a lot, what do you do about that? What is the switchover point between a "little" and a "lot"? Let us say you are in the middle of a rendezvous with a maneuver coming up and you have three sources of data each telling you what the maneuver should be. Also, your friend in the other spacecraft has a solution, as do the people on the ground. How do you choose among these solutions, if they differ? It is specific questions like these — and there are literally hundreds of them — that the mission-technique development process was set up to answer to everyone's satisfaction. Let me describe the development process and its products.

First of all, the task involves pinning down precisely how well the systems must work or if they are even needed to achieve mission success or to assure crew safety. This can become a tough job. Certainly, we do not want to abort a mission if it is safe to continue, but we must be sure it is safe, even if something else fails. To some extent, the decisions depend on where you are in the mission.

For example, we had as an Apollo 11 mission rule that we would not separate the lunar module (LM) from the command and service module (CSM) in lunar orbit if the rendezvous radar was not working. We felt that our lunar-orbit experience and LM systems maturity were not adequate at that time to start intentionally a rendezvous situation without the rendezvous radar. On the other hand, after the two spacecraft had been separated and descent had started, there was no reason to terminate the mission because of a rendezvous-radar failure. By that time, we would already be committed ourselves to performing the rendezvous without it and so might as well press on. Thus, some mission techniques can be chosen by applying common sense to the situation. Table 7-I illustrates one of the mission-technique products dealing with this kind of situation by listing which pieces of equipment must be working to continue at several go/no-go points in lunar orbit.

If a piece of equipment is obviously broken, it is easy to apply the mission rules agreed to before flight, such as those given in table 7-I. However, if it is just not working up to par, then what? In this case, the go/no-go decision must reflect system requirements in terms of the mission phase.

A good example of this was the preparation for the LM descent to the lunar surface on Apollo 11. Although all three gyroscopes in the primary guidance system inertial platform have identical design, the performance required of each was markedly different. Analysis showed that a misalinement about the pitch axis as large as 1° does not degrade the descent guidance unacceptably as long as the landing radar is working, but

TABLE 7-I. - MANDATORY GUIDANCE, NAVIGATION, AND CONTROL SYSTEMS

(a) Lunar module systems

LM systems	Undocking and separation[a]	Descent orbit insertion	Power descent initiate
Primary GNC system:			
LM guidance computer	R[b]	R	R
Inertial measurement unit	R	R	R
Display and keyboard	R	R	R
Abort guidance system	R	R	R
Control electronics system	R	R	R
Descent propulsion subsystem	R	R	R
Rendezvous radar	R	R	NR[c]
Landing radar	R	R	R
Flight director attitude indicator	R[d]	R[d]	R[d]
Alinement optical telescope	R	NR[e]	NR
Hand controllers[f]	R	R	R
Flashing light on LM	NR	R	R

[a]The separation maneuver and mini-football activities will be performed for all conditions allowing undocking.
[b]Required.
[c]Not required.
[d]Only one unit is required.
[e]Alinement optical telescope is required until the pre-descent-orbit-insertion fine alinement is completed.
[f]Translation and at least one rotation hand controller.

TABLE 7-I.- MANDATORY GUIDANCE, NAVIGATION, AND CONTROL SYSTEMS - Concluded

(b) Command and service module systems

CSM systems	Undocking and separation[a]	Descent orbit insertion	Power descent initiate
GNC system:			
CM computer	R[b]	R	NR[c]
Inertial measurement unit	R	R	NR
Display and keyboard	R	R	NR
Optics:			
Sextant	R	R	NR
Scanning telescope	R	R	NR
Crewman optical alinement sight	R	R	NR
Stabilization and control system:			
Body-mounted attitude gyroscopes	R[d]	R[d]	NR
Gyroscope display coupler	R	R	NR
Flight director attitude indicator	R[e]	R[e]	NR
Service propulsion system	R[f]	R[f]	NR
Hand controllers	R	R	NR
Entry monitor system change-in-velocity counters	NR	NR	NR
Very-high-frequency ranging	NR	NR	NR
Rendezvous radar transponder	R	R	NR

[a] The separation maneuver and mini-football activities will be performed for all conditions allowing undocking.

[b] Required.

[c] Not required.

[d] One set of body-mounted attitude gyroscopes required.

[e] Only one unit is required.

[f] If service propulsion system has failed, transearth injection will be performed at the next opportunity.

that just a 0.5° misalinement renders the guidance system incapable of performing a safe abort from the descent trajectory. So we established safe-abort ability as the criterion for computing the limiting pitch-axis gyroscope performance. Specifically, since the system is alined approximately an hour and a half before powered descent, we were able to fix 0.33 deg/hr as the maximum acceptable gyroscope pitch rate (the 3σ Apollo gyroscope performance is 0.1 deg/hr).

On the Apollo 11 mission, very large misalinements could be tolerated in yaw and roll. In fact, there was no reason not to continue unless one of these two gyroscopes drifted more than 1.5 deg/hr — the value determined by the gyroscope experts as an indication that the system is broken. In other words, a failure limit imposed a tighter constraint than any performance requirement; therefore, failure became the yaw and roll gyroscope criterion.

This will not be the case on Apollo 13, where the landing will take place in extremely rough terrain with a limited safe-landing area. Both mission success and crew safety depend on the guidance system getting the LM to within 1 kilometer of the aiming point. Accordingly, it is necessary to reduce the maximum allowable drift rate about the inertially vertical axis from the failure-criteria value of 1.5 to 0.145 deg/hr (i.e., 10 times better performance than was needed on the earlier missions). The job in each case is to find the most constraining criterion and the limits associated with it.

Once we set limits like these, we can then define precise crew and ground-support procedures for verifying that the performance comes within the limits before reaching each mission go/no-go decision point. It is also necessary to decide what to do if the performance does not fall between the limits. In the nonfailure cases just mentioned, it is possible to reload new drift-compensation values into the spacecraft computer by a command up link from the ground. In the failure case (i.e., 1.5-deg/hr drift on any of the three gyroscopes), the only choice is to "no-go" the descent.

Mission techniques also involve defining the procedures for monitoring systems during critical mission phases; for example, in powered flight, where an unsafe situation can develop very quickly. The way of doing this follows the pattern described for system-performance tests.

To illustrate, I have pulled the flow chart shown in figure 7-1 from one of our documents. It describes part of the decision logic followed by the ground flight controllers who monitor the lunar descent. The flow chart shows the steps to be taken under each condition in monitoring the guidance and control systems. The diagram is not meant to be followed step by step, but rather is a guide for the flight controllers, who monitor various parameters on strip charts. The flow chart gives the procedures to be followed if certain limits are exceeded.

Monitoring another type of critical phase must also be worked out. During lunar-orbit rendezvous, we have no fewer than five first-class systems computing the rendezvous maneuvers. The LM has its primary guidance system and a backup guidance system; the CSM has an excellent system, and the ground can do a very good job too. In addition to those, the crew can do the same job with some charts and simple paper-and-pencil calculations. The rendezvous is probably the best example of where a job we call "data priority" had to be applied. It is clear how the maneuvers should be executed if all of these data sources agree with one another, but how should you respond

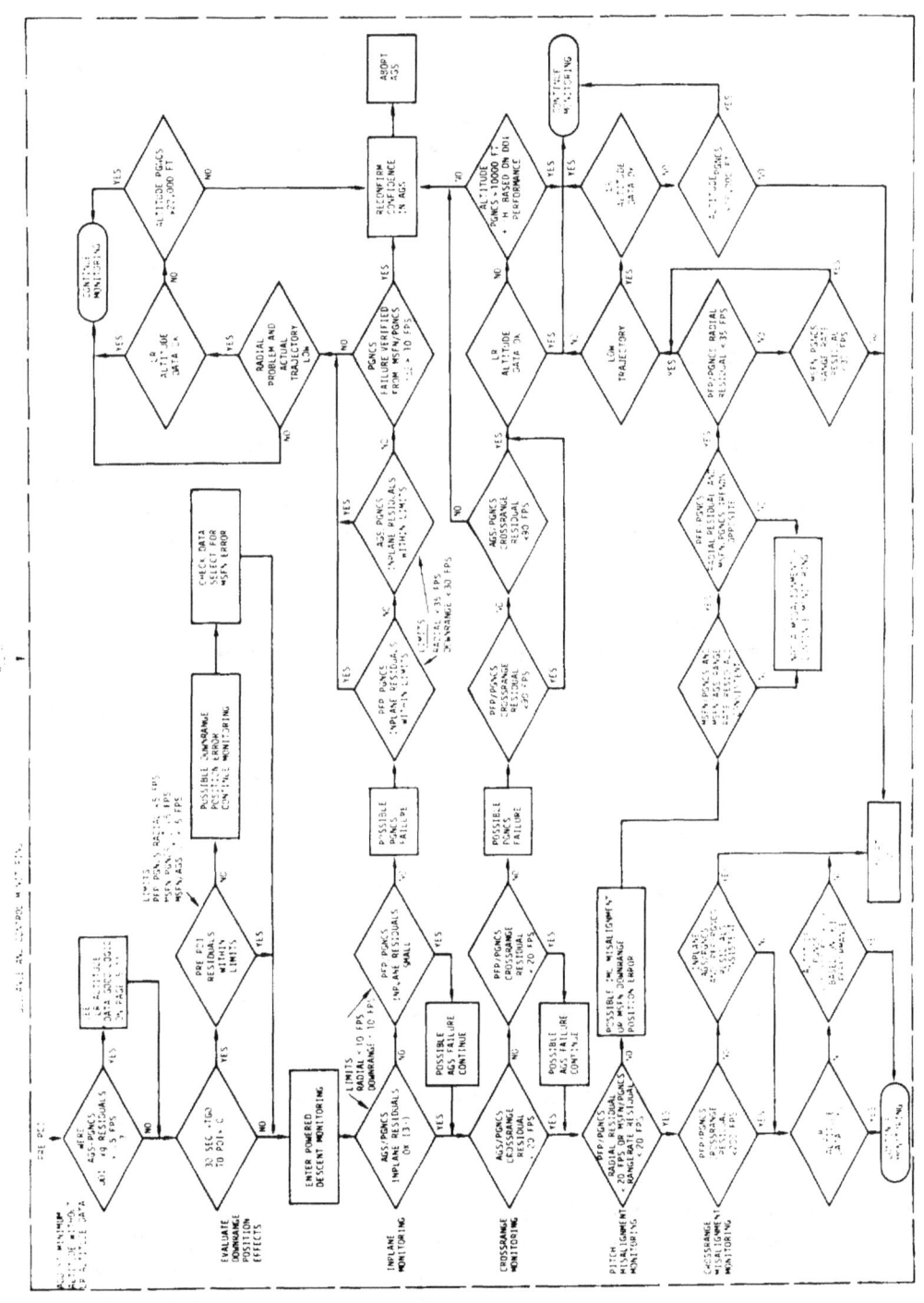

Figure 7-1.— Steps the ground-based flight controllers take if certain guidance and control values exceed premission limits for the LM descent to the lunar surface.

if they do not? Many hours of discussion were spent in pinning down such basics as what we mean by "agreement" (e.g., within 0.1, 1.0, or 10 fps for velocity) and, if things do not agree, which one should be used?

We also produce mission-technique documents for every phase of every manned Apollo flight. The mission-technique documents describe precisely how all of this is done and the reasons for doing it that way. The documents are widely distributed to make sure everyone knows what we plan to do. But most important, all of these techniques are included in Crew Procedures, Checklist, Flight Plan, Flight Control Procedures, and Mission Rules — the documents which truly govern the conduct of the mission.

Development of mission techniques was achieved by assigning to an individual in the Apollo Spacecraft Program Office the responsibility of coordinating the activities of the various groups working on those things so that when they were finished, there was some assurance they would all be compatible, complete, and universally understood.

Of course, the way we got this job done was with meetings — big meetings, little meetings, hundreds of meetings! The thing we always tried to do in these meetings was to encourage everyone, no matter how shy, to speak out, hopefully (but not always) without being subjected to ridicule. We wanted to make sure we had not overlooked any legitimate input.

One thing we found to be very effective — and which we almost always did — was to make decisions on how to do the job, even if the data available were incomplete or conflicting, or if there was substantial disagreement among the participants. This even included making educated guesses at the performance abort limits and stating that they were the values we would use unless someone came in with something better. However, do not get me wrong. These limits are a very serious business. They literally define the point at which a mission will be aborted. You can imagine how emotional our meetings frequently were!

Although the decisions reached often displeased someone, the fact that a decision had been made was invaluable. Since this effort was officially sanctioned, the decisions served to unify all subsequent work. They often also pointed up the need for more work. And for those dissatisfied with the decisions, at least they presented a firm target which they could attack through recognized channels to higher authority.

In short, the primary purpose of these meetings was to make decisions, and we never hesitated! These early decisions provided a point of departure; and by the time the flight took place, our numbers were firm, checked, and double-checked. By then, we knew they were right!

Mission-Technique Spinoffs: There were several important spinoffs from this work. The meetings were regularly attended by experts involved in all facets of trajectory control — systems, computer, and operations people, including the crew. Our discussions not only resulted in agreement among everyone as to how we planned to do the job and why, but also inevitably educated everyone as to precisely how the systems themselves work, down to the last detail. A characteristic of Apollo you could not help noting was just how great the lack of detailed and absolute comprehension are on a program of this magnitude. There is a basic communication problem for which I can offer no acceptable solution. To do our job, we needed a level of detailed understanding of the functioning of systems and software far greater than was generally available. Through our meetings, however, we forced this understanding. It was not easy, but we got it sorted out eventually — together.

Another valuable spinoff from these discussions was the state of readiness achieved by the operations people — ground and crew — before starting simulation training exercises. By the time we had finished our work, everyone had a pretty good idea of exactly what we were trying to do and how we were going to do it — exactly what was expected of each individual and when. And, most important, the data flow between ground and spacecraft would have been defined and scheduled in detail. As a result, when simulations began, they could proceed with maximum efficiency doing exactly what they were intended to do — train people. Interruptions and false starts to get things squared away were kept to a minimum, usually just involving those things that had not occurred to us previously or which did not work out too well in actual practice. I cannot overemphasize the importance of this, since obtaining adequate training is one of the toughest jobs we have.

A third spinoff involves the inevitable discovery of system deficiencies. When you get a gang of people like these together to figure out exactly how you are going to do the trajectory-control job, it is inevitable that really outstanding new ideas will emerge. These new ideas caused a large number of spacecraft and control-center computer-program changes, both additions and — I am pleased to report — deletions. We also uncovered undesirable or even unacceptable hardware characteristics. We were able to get some fixed. We took it upon ourselves to advertise the rest extensively and to plan workaround procedures.

The greatest impact on an area, outside of mission techniques, involved trajectory planning. As we worked out the details, it was found advisable to make substantial changes in lunar-orbit rendezvous, lunar descent aborts, and entry through the earth atmosphere, as well as many less significant things, such as performing the lunar-orbit insertion in two steps instead of one and, starting with Apollo 13, doing the descent-orbit insertion maneuver with the CSM instead of the LM.

These changes were usually made when it was found that suitable monitoring techniques were not available or that it was necessary to make the mission more compatible with the backup techniques. For example, there are two levels of backup to the primary G&N system of the CSM for atmospheric entry at the conclusion of a lunar mission. However, neither of the backups can support the long-range entry involving skipping out of the atmosphere, which was originally planned for Apollo missions to avoid bad weather and was the standard mode the primary guidance system was designed to fly. Accordingly, to permit switchover of systems if the primary G&N failed during entry, the trajectory plan was changed and the G&N was modified to fly the nonskip trajectory we now use on all lunar flights.

How has this all worked out now that we have essentially finished it? Well, I have to confess that, as far as I know, there was never an instance in any of the Apollo flights where the detailed system test and monitoring we set up uncovered anything unacceptable. That is, we could have flown with our eyes closed, and the primary guidance system would have come through. The work we were originally set up to do has never saved us from catastrophe, although I am sure it must have reduced anxieties by several orders of magnitude. As far as the usefulness of the overall effort, I really do not see how we could have flown the missions without its having been done. The spin-offs alone were worth it.

8. FLEXIBLE YET DISCIPLINED MISSION PLANNING

By C. C. Kraft, Jr., J. P. Mayer, C. R. Huss, and R. P. Parten
Manned Spacecraft Center

Apollo mission planning extended over a period of approximately 8 years and encompassed many technical disciplines. It progressed from the early design-reference lunar mission in 1962 through the detailed operational planning of specific missions which began in 1964. A pattern of mission-planning procedures, techniques, and management processes evolved that brought order to the many (and sometimes diverse) elements of the mission-planning team. This report examines the Apollo mission-planning process, the important considerations that influenced the mission planners, and the evolution of the Apollo development flight schedule. The report concludes with a brief description of the more important panels, meetings, and working groups that helped to coordinate the mission-planning activities.

The planning and design of a mission, like the development of spacecraft hardware, proceeds from previously specified objectives and becomes constrained by system characteristics and operational considerations. Fundamentally, the process consists of a series of iterative cycles (fig. 8-1) in which a design is defined to increasingly finer levels of detail as the program progresses and as the flight hardware and operational considerations become better known. Initially, mission design has the purpose of transforming broad objectives into a standard profile and sequence of events against which the space-vehicle systems can be designed. Usually, incompatibilities arise immediately between system design and mission design, and these necessitate trade-off studies to arrive at a compromise. Later, as hardware design solidifies, the emphasis of mission design becomes more operationally oriented, as the mission planners attempt to constrain their design to the operating capabilities of the space vehicles, ground-support facilities, and flight software. The final planning phase, which occurs during the year before a launch, involves development of the detailed procedures, techniques, and mission rules which are used by the flight crew and ground control team for both nominal and contingency missions.

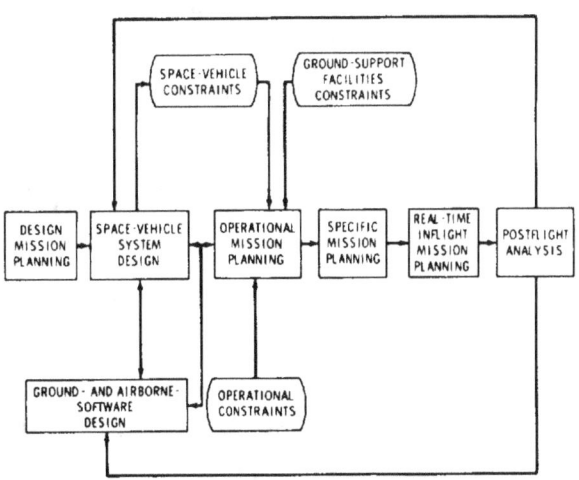

Figure 8-1.- Iterative mission-planning process.

MAJOR CONSIDERATIONS

The considerations important to Apollo mission planners depended somewhat on the type of mission: manned or unmanned, development or operational. For manned missions, the prime consideration was crew safety, whereas unmanned missions stressed mission success. For development missions, the prime consideration was to maximize the number of test objectives that could be met successfully, as opposed to operational missions, in which the number of final mission objectives was maximized.

The need for early decisions made the Apollo mission planners the driving force in stating the requirements for collecting and documenting all constraints. This need existed particularly during the development phase of the program. After the constraints were obtained, the development of mission techniques that would achieve the required test or mission objectives within them was necessary. This development proved to be an iterative process which involved constraints, techniques, and objectives. It usually was carried to the time of final mission rules and flight plans. The iteration process and the consideration of possible failures or contingency situations entailed much alternate mission and contingency mission planning. This type of planning activity reached a peak during the manned development missions.

Because of involvement early in the planning stage of a mission, the mission planners have been able to affect materially the development of both onboard and ground software. In many cases, the software had to be tailored to a specific mission, although every effort was made to use the final program software wherever possible.

The software development schedule and costs can (and did) influence the design of some missions. In the early days of Apollo, separate programs were being developed for the unmanned Saturn IB and Saturn V test missions, because of distinct differences in the test requirements for the missions. Later, it became apparent that the development of several different spacecraft- and ground-software programs was not feasible, considering the schedule and costs for delivering the lunar-landing-mission program. Because of this, the Saturn V mission test objectives were reexamined and were modified to accept the software used for the Saturn IB missions.

OPERATIONAL PLANNING CYCLE

Usually, the Apollo operational planning cycle began between 36 and 18 months before a launch, depending on the extent of mission complexity. Because several organizations within NASA and throughout industry were participating in the analysis and design of the Apollo development and lunar missions, a need was identified for coordinating the efforts of all mission-planning elements.

In late 1964, an Apollo Mission Trajectory Documentation Plan was prepared, which defined the principal documents and information flow in the areas of mission planning and trajectory analysis. This plan became a major vehicle for coordinating the more significant mission-planning milestones throughout the Apollo Program.

The operational mission design was divided into three separate phases: mission definition, mission design, and flight preparation. Schedules for the beginning and termination of each phase were established specifically to dovetail with the major hardware development, airborne- and ground-software development, and crew-training milestones.

Although each phase sometimes specified as many as 20 major milestones (documents), the nucleus of the plan involved only four basic milestones for each phase: a mission requirements document, an operational data book, a trajectory plan, and of course a flight plan. The remainder of the milestones concerned abort and contingency planning, dispersion analyses, consumables analyses, and other special requirements, such as range-safety plans, orbital-debris studies, onboard-data-file information, and crew-simulator data.

Early in the Apollo Program, considerable difficulty was experienced in the exchange, standardization, and dissemination of critical data required both by and from the mission planners. In view of the interdependency of most mission-planning milestone documents, the need for coordination and tight control in disseminating these data became acute. As a result, the Apollo Spacecraft Program Office in conjunction with the hardware contractors and operational elements exercised strict controls and procedures in governing mission-planning data. As a part of this program, most organizations appointed key personnel on a full-time basis to support the data-management network. In retrospect, this action must be regarded as vital in consolidating and strengthening the Apollo mission-planning process.

The next step in the Apollo operational planning cycle involves what came to be known as mission techniques, which were developed in the form of logic flows that detail each decision point, threshold value, and ground rule for each phase of both nominal and contingency missions. Section 7 discusses mission techniques in some detail.

In Apollo, as any other complex space mission, it is virtually impossible to develop premission plans for every contingency that could arise during flight. Although specific plans are developed for all abort potentialities that involve crew safety, most alternate missions are developed on a class basis by using the alternate test and mission objectives. However, the real-time mission planner is given a powerful assortment of mission-planning computer programs that enhance his ability to manage any contingency. By the proper use of these on-line computer programs, alternate mission plans can be developed in real time and can thereby augment the premission planning activities.

DEVELOPMENT FLIGHT SCHEDULE

To understand the complexity of the task of mission planning for the Apollo Program, the evolvement of the development flight schedule may be reviewed. From the beginning of the Apollo Program, it was recognized that the flight tests for verifying hardware design, hardware performance, and operational techniques would be extensive. At the same time, it was recognized that the cost and scheduling of such flights must be controlled and limited to keep development costs for the program at a minimum.

Test requirements for the major hardware elements (launch-escape system, command module aerodynamics, spacecraft structural verification, thermal-protection system, separation systems, communications systems, propulsive systems, landing systems, etc.) were recognized early as tests that could be made by unmanned flight vehicles. This realization resulted in the series of Little Joe II test flights (launched at the NASA White Sands Test Facility) and in three Saturn IB and two Saturn V test flights (launched at the NASA John F. Kennedy Space Center), which can be noted in table 8-I.

TABLE 8-I. - APOLLO SPACECRAFT FLIGHT HISTORY

Mission	Spacecraft	Description	Launch date	Launch site
PA-1	BP-6	First pad abort	Nov. 7, 1963	White Sands Missile Range, N. Mex.
A-001	BP-12	Transonic abort	May 13, 1964	White Sands Missile Range, N. Mex.
AS-101	BP-13	Nominal launch and exit environment	May 28, 1964	Kennedy Space Center, Fla.
AS-102	BP-15	Nominal launch and exit environment	Sept. 18, 1964	Kennedy Space Center, Fla.
AS-002	BP-23	Maximum dynamic pressure abort	Dec. 8, 1964	White Sands Missile Range, N. Mex.
AS-103	BP-16	Micrometeoroid experiment	Feb. 16, 1965	Kennedy Space Center, Fla.
A-003	BP-22	Low-altitude abort (planned high-altitude abort)	May 19, 1965	White Sands Missile Range, N. Mex.
AS-104	BP-26	Micrometeoroid experiment and service module reaction control system launch environment	May 25, 1965	Kennedy Space Center, Fla.
PA-2	BP-23A	Second pad abort	June 29, 1965	White Sands Missile Range, N. Mex.
AS-105	BP-9A	Micrometeoroid experiment and service module reaction control system launch environment	July 30, 1965	Kennedy Space Center, Fla.
A-004	SC-002	Power-on tumbling boundary abort	Jan. 20, 1966	White Sands Missile Range, N. Mex.
AS-201	SC-009	Supercircular entry with high heat rate	Feb. 26, 1966	Kennedy Space Center, Fla.
AS-202	SC-011	Supercircular entry with high heat load	Aug. 25, 1966	Kennedy Space Center, Fla.
Apollo 4	SC-017 LTA-10R	Supercircular entry at lunar return velocity	Nov. 9, 1967	Kennedy Space Center, Fla.
Apollo 5	LM-1	First lunar module flight	Jan. 22, 1968	Kennedy Space Center, Fla.
Apollo 6	SC-020	Verification of closed-loop emergency detection system	April 4, 1968	Kennedy Space Center, Fla.
Apollo 7	CSM 101	First manned flight; earth orbital	Oct. 11, 1968	Kennedy Space Center, Fla.
Apollo 8	CSM 103	First manned lunar orbital flight; first manned Saturn V launch	Dec. 21, 1968	Kennedy Space Center, Fla.
Apollo 9	CSM 104 LM-3	First manned lunar module flight; earth orbit rendezvous; extravehicular activity	March 3, 1969	Kennedy Space Center, Fla.
Apollo 10	CSM 106 LM-4	First lunar orbit rendezvous; low pass over lunar surface	May 18, 1969	Kennedy Space Center, Fla.
Apollo 11	CSM 107 LM-5	First lunar landing	July 16, 1969	Kennedy Space Center, Fla.
Apollo 12	CSM 108 LM-6	Second lunar landing	Nov. 14, 1969	Kennedy Space Center, Fla.

The Little Joe II flights were unmanned ballistic flights strictly for command and service module (CSM) hardware development, mainly of the launch-abort escape system, separation systems, spacecraft aerodynamics and structural integrity, and landing systems. The Saturn IB series extended these tests to more extreme conditions (higher entry speeds, higher heating conditions, longer flight times). Also, the Saturn IB flights began testing the propulsion and guidance and control systems, on both the CSM and the landing module. The Saturn V flights completed the unmanned test program, which exercised the CSM at very near the conditions expected for a lunar mission, the major variance being less flight time. By their nature, these missions required that many special hardware and software elements be developed which were not applicable to the manned program. However, they did aid considerably in the development of ground operational techniques, mission rules, spacecraft checkout, and launch preparation.

Many systems, because of their complexity and numerous modes of operation (electrical power, environmental, computer, communications) could be tested only on manned missions. The number of missions being planned was quite large in the early days of the program, not because of the spacecraft test requirements, but because of the uncertainty of success of the development program, both from the spacecraft and launch-vehicle standpoint. In late 1963, there were eight Saturn IB and eight Saturn V development missions scheduled and in the planning stages, and approximately six to eight more backup missions in the planning stages.

During the period from late 1967 to early 1968, a major redefinition of the development flight program reduced to a minimum the number of development flights leading to the landing mission. After the first two Saturn IB flights, the program had been reduced to two Saturn IB and seven Saturn V development missions, and only two Saturn IB backup missions were being planned. This reduction reflected successes with Little Joe II, the two Saturn IB missions, and Saturn development tests, as well as increased confidence in the system performance and reliability bases on flight and ground tests. In mid-1968, after the successful unmanned Saturn V missions, a partially successful unmanned lunar module mission, and a successful manned CSM mission, the development flight test program was reduced to three Saturn V missions with complete spacecraft: a low-earth-orbit mission (D), a high-ellipse earth-orbit mission (E), and a lunar-orbit mission (F).

Because of the high confidence in the CSM, which was established by the early missions, and because of the requirement of additional checkout and testing of the first manned lunar module, it was decided to fly a CSM alone on a lunar-orbit mission (C). This mission provided an early evaluation of lunar navigation techniques and operational procedures, and added immeasurably to the progress and to the eventual success of the program. It was also decided at this time to eliminate the high-ellipse earth-orbit mission (E) in favor of doing tests in the actual lunar environment. Of course, these extremely complicated E missions completed the necessary testing and development of hardware, software and operating techniques, ground checkout and launch preparation, and development of mission rules and overall operational capability, except for the actual lunar landing. This phase of the mission could be tested only through the use of the lunar landing training vehicle and by the actual lunar landing.

MISSION-PLANNING COORDINATION

Because the development of the Apollo mission plans involved close cooperation of three widely separated NASA centers, intercenter panels were organized to coordinate all activities in which an interface occurred between the launch vehicle and the spacecraft. For example, early in the program both the launch vehicle and the spacecraft had development objectives which had to be met with as few flights as possible. The development of missions which supported adequately the mission objectives of both launch vehicles and spacecraft took considerable planning effort among the centers.

At the NASA Manned Spacecraft Center (MSC), a series of flight operations plans meetings was conducted to develop the basic mission plan after the issuance of requirements. These meetings brought together the various experts in spacecraft systems, trajectory analysis, and guidance and control and the flight controllers, crew, and flight plan developers from contractors and MSC organizations. (See fig. 8-2.) The chief importance of these meetings was to set constraints on the mission.

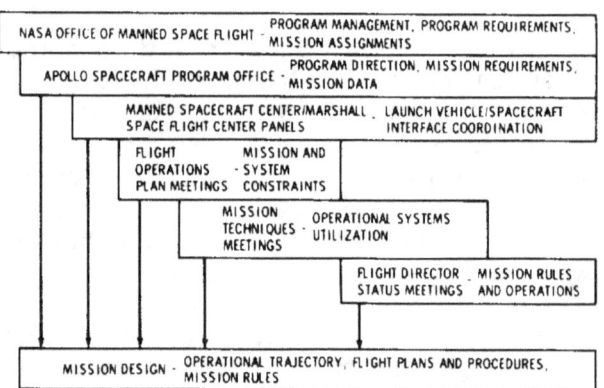

Figure 8-2. - Apollo mission design instrumentation.

Mission techniques meetings took over where the flight operations meetings left off. The flight operations meetings led to a definition of the basic mission and the mission constraints. The detailed operational procedures for flying the mission were developed in the mission techniques meetings. The basic mission-planning documents which these working groups influenced were the operational trajectory, the flight plan and procedures, and the mission rules.

As changes to hardware must be controlled, so must changes to mission plans. Mission plan changes were controlled by three basic control boards. The Apollo Spacecraft Configuration Control Board (directed by the Manager of the Apollo Spacecraft Program) exerted control over all changes that affected mission objectives, hardware, trajectories, and propellant requirements. The Software Configuration Control Board (directed by the Director of Flight Operations) controlled all changes to the onboard and ground computing programs. The Crew Procedures Configuration Control Board (directed by the Director of Flight Crew Operations) controlled changes to the mission flight plan and all operational crew procedures. The Data and Requirements Control Panel controlled operational data affecting the mission flight plan.

CONCLUDING REMARKS

In planning the Apollo missions, much emphasis was placed on the demand for flexibility in the development program and responsiveness to changing needs. The dynamic conditions present in Apollo strongly influenced the mission planners in providing comprehensive alternate mission capability and flexibility in the ground and airborne flight software. Probably of more importance, however, was the capacity of the mission-planning team to react to major program readjustments, as evidenced typically by the Apollo 8 success. The effectiveness of this team, by using the process described here, is measured by the Apollo record.

www.ingramcontent.com/pod-product-compliance
Lightning Source LLC
Chambersburg PA
CBHW081735170526
45167CB00009B/3823